BEYOND THE MARKET

Designing Nonmarket Accounts for the United States

Panel to Study the Design of Nonmarket Accounts

Katharine G. Abraham and Christopher Mackie, Editors

Committee on National Statistics

Division of Behavioral and Social Sciences and Education

NATIONAL RESEARCH COUNCIL
OF THE NATIONAL ACADEMIES

THE NATIONAL ACADEMIES PRESS
Washington, D.C.
www.nap.edu

2005

THE NATIONAL ACADEMIES PRESS 500 Fifth Street, NW Washington, DC 20001

NOTICE: The project that is the subject of this report was approved by the Governing Board of the National Research Council, whose members are drawn from the councils of the National Academy of Sciences, the National Academy of Engineering, and the Institute of Medicine. The members of the committee responsible for the report were chosen for their special competences and with regard for appropriate balance.

This study was supported by an unnumbered contract between the National Academy of Sciences and Yale University and the Glaser Family Foundation. Support of the work of the Committee on National Statistics is provided by a consortium of federal agencies through a grant from the National Science Foundation (Number SBR-0112521). Any opinions, findings, conclusions, or recommendations expressed in this publication are those of the author(s) and do not necessarily reflect the views of the organizations or agencies that provided support for the project.

Library of Congress Cataloging-in-Publication Data

Beyond the market : designing nonmarket accounts for the United States / Katharine G. Abraham and Christopher Mackie, editors.

p. cm.

Includes bibliographical references.

ISBN 0-309-09319-8 (pbk.)—ISBN 0-309-54592-7 (pdf)

1. Accounting—United States. 2. Social accounting—United States. 3. National income—Accounting. I. Abraham, Katharine G. II. Mackie, Christopher D.

HF5616.U5B473 2005

339.373—dc22

2004026247

Additional copies of this report are available from The National Academies Press, 500 Fifth Street, NW, Lockbox 285, Washington, DC 20055; (800) 624-6242 or (202) 334-3313 (in the Washington metropolitan area); Internet, http://www.nap.edu.

Printed in the United States of America.

Suggested citation: National Research Council. (2005). *Beyond the Market: Designing Nonmarket Accounts for the United States*. Panel to Study the Design of Nonmarket Accounts, K.G. Abraham and C. Mackie, eds. Committee on National Statistics, Division of Behavioral and Social Sciences and Education. Washington, DC: The National Academies Press.

THE NATIONAL ACADEMIES

Advisers to the Nation on Science, Engineering, and Medicine

The **National Academy of Sciences** is a private, nonprofit, self-perpetuating society of distinguished scholars engaged in scientific and engineering research, dedicated to the furtherance of science and technology and to their use for the general welfare. Upon the authority of the charter granted to it by the Congress in 1863, the Academy has a mandate that requires it to advise the federal government on scientific and technical matters. Dr. Bruce M. Alberts is president of the National Academy of Sciences.

The **National Academy of Engineering** was established in 1964, under the charter of the National Academy of Sciences, as a parallel organization of outstanding engineers. It is autonomous in its administration and in the selection of its members, sharing with the National Academy of Sciences the responsibility for advising the federal government. The National Academy of Engineering also sponsors engineering programs aimed at meeting national needs, encourages education and research, and recognizes the superior achievements of engineers. Dr. Wm. A. Wulf is president of the National Academy of Engineering.

The **Institute of Medicine** was established in 1970 by the National Academy of Sciences to secure the services of eminent members of appropriate professions in the examination of policy matters pertaining to the health of the public. The Institute acts under the responsibility given to the National Academy of Sciences by its congressional charter to be an adviser to the federal government and, upon its own initiative, to identify issues of medical care, research, and education. Dr. Harvey V. Fineberg is president of the Institute of Medicine.

The **National Research Council** was organized by the National Academy of Sciences in 1916 to associate the broad community of science and technology with the Academy's purposes of furthering knowledge and advising the federal government. Functioning in accordance with general policies determined by the Academy, the Council has become the principal operating agency of both the National Academy of Sciences and the National Academy of Engineering in providing services to the government, the public, and the scientific and engineering communities. The Council is administered jointly by both Academies and the Institute of Medicine. Dr. Bruce M. Alberts and Dr. Wm. A. Wulf are chair and vice chair, respectively, of the National Research Council.

www.national-academies.org

PANEL TO STUDY THE DESIGN OF NONMARKET ACCOUNTS

KATHARINE G. ABRAHAM *(Chair)*, Joint Program in Survey Methodology, University of Maryland
DAVID CUTLER, Department of Economics, Harvard University
NANCY FOLBRE, Department of Economics, University of Massachusetts, Amherst
BARBARA FRAUMENI, Bureau of Economic Analysis, U.S. Department of Commerce, Washington, DC
ROBERT E. HALL, Hoover Institution, Stanford University
DANIEL S. HAMERMESH, Department of Economics, University of Texas, Austin
ALAN B. KRUEGER, Woodrow Wilson School, Princeton University
ROBERT MICHAEL, Harris School of Public Policy, University of Chicago
HENRY M. PESKIN, Edgevale Associates, Nellysford, VA
MATTHEW D. SHAPIRO, Department of Economics, University of Michigan
BURTON A. WEISBROD, Department of Economics, Northwestern University

CHRISTOPHER MACKIE, *Study Director*
MICHAEL SIRI, *Senior Program Assistant*
MARISA GERSTEIN, *Research Assistant*

COMMITTEE ON NATIONAL STATISTICS
2004

Preface

One of the long-standing goals of the Committee on National Statistics (CNSTAT) is the improvement of economic measurement and the data sources crucial to that measurement. In working toward that goal, recent CNSTAT panels have produced reports on price and cost-of-living indexes, poverty measurement, measurement of the economy's government sector, and the design of environmental and natural resource accounts. The last report in this list, *Nature's Numbers,* focused on goods and services associated with the environment, which are in many cases not transacted in markets and hence not captured in conventional economic accounts. That report did much to set the conceptual stage for this panel's broader study of economic activities that are largely nonmarket in character.

This report is the product of contributions from many individuals. The project was sponsored by the Yale University Program on Nonmarket Accounts which, in turn, was funded by a grant from the Glaser Foundation. The Yale program is directed by William Nordhaus, whose long history of pioneering research in this and related areas—dating back three decades to his work with James Tobin on measures of economic welfare and continuing through his chairing of the *Nature's Numbers* panel—helped to establish the foundations for this panel's work. Dr. Nordhaus, along with Martin Collier of the Glaser Foundation, attended the first meeting and, in articulating their hopes for the study, helped the panel sharpen its vision of their charge. The panel is grateful also to Dan Melnick who served as liaison to the panel for the Yale Program and contributed valuable suggestions and points of clarification along the way.

Many others generously presented material at panel meetings and answered questions from panel members, thereby helping us to develop a broader and deeper understanding of key methodological and data issues relevant to the construction of nonmarket accounts. The panel especially thanks Steven Landefeld, director of the Bureau of Economic Analysis, who provided insights based on his long experience and extensive knowledge of economic accounting; Diane Herz and Lisa Schwartz, of the Bureau of Labor Statistics, who educated the panel about that agency's important new time-use survey; Thomas Juster, University of Michigan, and Robert Pollak, Washington University, who shared their expertise on conceptual and measurement issues relating to time use and the theory of time allocation; Suzanne Bianchi, who provided tabulations of time-use data and information about the underlying surveys; and Peter Harper, Australian Bureau of Statistics, and Sue Holloway, Office for National Statistics, United Kingdom, who informed the panel about some of the exciting work on nonmarket accounting underway in other countries.

The meetings of the panel also provided many opportunities for the panel members to learn from one another. Each of the panel members contributed indispensable special expertise to the preparation of the panel's final report. Katharine Abraham and Robert Hall wrote the first draft of the report's introduction; Barbara Fraumeni prepared a description of the existing national accounts that makes up part of Chapter 2; Daniel Hamermesh contributed a description of the new American Time Use Survey that also appears in Chapter 2; Nancy Folbre and Daniel Hamermesh prepared the first draft of Chapter 3, on the topic of household production; Nancy Folbre and Robert Michael wrote the initial draft of the material on the role of families in the production of human capital that eventually found its way into Chapter 4 of the panel's final report; Barbara Fraumeni and Alan Krueger took the lead on the preparation of Chapter 5, on accounting for investments in education; David Cutler and Matthew Shapiro provided a first draft of Chapter 6, on accounting for investments in health; and Henry Peskin and Burton Weisbrod worked together on the initial drafts of Chapters 7 and 8, on accounting for the activities of nonprofits and governments and accounting for the environment. All of the report's chapters underwent several rounds of significant revision, reflecting intensive discussion and debate that involved the full panel, but these productive exchanges could not have occurred had individual panel members not taken the lead in preparing the first drafts that served as our starting point.

A special comment is needed about one of our panel members, Barbara Fraumeni, who is the chief economist at the Bureau of Economic Analysis (BEA). Although government employees who are technical experts in their fields may serve on study panels for the National Academies, precautions are taken in such cases to ensure against real or perceived conflicts of interest. In this case, the institution recognized both that this panel might make recommendations directly related to the work of the BEA and that Dr. Fraumeni's unrivaled expertise on national economic accounting in general and the U.S. National Income and Product Accounts in particular would be critical to the panel's work. After careful consideration of these factors, the institution invited Dr. Fraumeni to serve on the panel.

We also note the contributions of two original panel members, Dora Costa, Massachusetts Institute of Technology, and Daniel Kahneman, Princeton University, who attended meetings early in the panel's $2^1/_2$ years of work and provided keen insights that helped the panel to chart its course. We are sorry that they were unable to continue as active members.

The panel could not have conducted its work without an excellent and well-managed staff. Andy White was the director of CNSTAT at the time the panel was formed, and we appreciate his support for the panel's work. Project assistants Michael Siri and Marisa Gerstein provided excellent administrative, editorial, and research support. The panel also benefited from the work of Eugenia Grohman and Kirsten Sampson Snyder, both of the Division of Behavioral and Social Sciences and Education, who were responsible for editing the report and overseeing the review process.

The entire panel owes a special debt of gratitude to Christopher Mackie, the panel's study director. During the course of the panel's deliberations, he played an invaluable role in facilitating communication among panel members, drawing the panel's attention to relevant studies that we might otherwise have overlooked, helping to develop the structure for the panel's final report, and directing the panel's attention to gaps and inconsistencies in the discussion of different topics that needed to be addressed. Over the past year, in collaboration with various panel members, he read and reworked each of the report's chapters multiple times, making improvements on each pass and helping to turn an initially disparate set of individual chapter drafts into a more integrated whole, and then shepherded the report through the final review process. For me personally, working with Chris was a great pleasure, and I know I speak for the entire panel in expressing my gratitude to him for his dedicated professionalism, reliable good cheer, and many substantive contributions.

The report has been reviewed in draft form by individuals chosen for their diverse perspectives and technical expertise, in accordance with procedures approved by the Report Review Committee of the National Research Council (NRC). The purpose of this independent review is to provide candid and critical comments that will assist the institution in making the published report as sound as possible and to ensure that the report meets institutional standards for objectivity, evidence, and responsiveness to the study charge. The review comments and draft manuscript remain confidential to protect the integrity of the deliberative process.

We thank the following individuals for their participation in the review of this report: John C. Bailar, III, Department of Health Studies (emeritus), University of Chicago; Robert Haveman, Department of Economics, University of Wisconsin-Madison; J. Steven Landefeld, Bureau of Economic Analysis, Washington, DC; Arleen Leibowitz, Department of Policy Studies, University of

California, Los Angeles; Robert A. Margo, Department of Economics and History, Vanderbilt University, and Research Associate, National Bureau of Economic Research; Timothy Smeeding, Center for Policy Research, Syracuse University; Frank P. Stafford, Department of Economics, University of Michigan; and Frances Woolley, Department of Economics, Carleton University, Ottawa, Ontario Canada.

Although the reviewers listed above provided many constructive comments and suggestions, they were not asked to endorse the conclusions or recommendations, nor did they see the final draft of the report before its release. The review of this report was overseen by Robert A. Pollak, Department of Economics, Washington University, St. Louis, MO, and Joseph P. Newhouse, School of Health Policy and Management, Harvard University. Appointed by the National Research Council, they were responsible for making certain that an independent examination of this report was carried out in accordance with institutional procedures and that all review comments were carefully considered. Responsibility for the final content of this report rests entirely with the authoring committee and the institution.

Katharine G. Abraham, *Chair*
Panel to Study the Design of Nonmarket Accounts

Contents

EXECUTIVE SUMMARY 1

1 INTRODUCTION 9
Background and Overview, 9
Motivation, 12
Priorities, Scope, and Objectives, 14
A Conceptual Framework, 23

2 ACCOUNTING AND DATA FOUNDATIONS 39
Overview of the National Income and Product Accounts, 40
Measuring Time Use, 43
Demographic Data, 52
Other Data Needs, 54

3 HOME PRODUCTION 55
The Household as a Factory, 59
Measuring Inputs, 63
Valuing Inputs, 68
Measuring and Valuing Output, 74
Data Requirements, 76

4 THE ROLE OF THE FAMILY IN THE PRODUCTION OF HUMAN CAPITAL 79
Conceptual Framework, 79
Defining Human Capital, 81
The Human Capital Production Function, 83
Family Inputs to the Development of Children's Human Capital, 88
Valuing the Time Parents Devote to Children, 91

5 EDUCATION 93
Conceptual Framework, 94
Measuring and Valuing Inputs, 97
Measuring and Valuing Output, 105
Other Issues, 116

6 HEALTH 117
Conceptual Framework, 119
Measuring and Valuing Inputs, 125
Measuring and Valuing Health, 131
Data Requirements, 140

7 THE GOVERNMENT AND PRIVATE NONPROFIT SECTORS 141
Conceptual Framework, 143
Volunteer Labor, 146
Donated Goods, 153
Measuring and Valuing Output, 153
Data Requirements, 159
Conclusions, 160

8 THE ENVIRONMENT 163
Definition and Scope of Coverage, 164
Current Accounting Approaches, 169
Future Directions, 171
A Note on the Social Environment, 175

REFERENCES 179

APPENDIX
Biographical Sketches of Panel Members and Staff 195

INDEX 201

BEYOND THE MARKET

Executive Summary

BACKGROUND AND PANEL CHARGE

National income and product accounts (NIPAs) are, in most countries, the data source for the most influential measures of overall economic activity. Key benchmarks derived from these accounts—most notably gross domestic product (GDP)—along with other economic data such as price and employment statistics, are widely viewed as indicators of how well a nation is doing.

Nevertheless, since their earliest construction for the United States by Simon Kuznets in the 1930s, concerns have been voiced that the accounts are incomplete. While broadly accepted and precisely estimated, the NIPAs omit a large part of the nation's product. They emphasize market transactions and reveal little about production in the home and other nonmarket settings. Furthermore, in some areas in which activity is organized in markets (e.g., medical care), the existing accounts do not capture key elements (e.g., the social value of medical innovations or unpaid time spent caring for the sick).

Nonmarket accounts would be particularly helpful in improving our understanding of the sources of growth in the economy. Researchers now must supplement data from the national accounts with external estimates of the contributions of research and development, natural resources, and investments in human capital. These limitations of national accounting data reflect the reality that neither economic production nor contributions to social welfare stop at the market's border, but extend to many nonmarket activities. Failure to account for these activities may significantly distort policy makers' sense of economic trends and the desirability of potential policy interventions.

Given that limitations of the NIPAs have long been recognized, why is this report needed at this time? Few would dispute the existence of economically valued nonmarket inputs and outputs. Examples are easy to find: the higher value of a house sold after home improvements are made by the homeowner, the development of attitudes and skills that have value in the marketplace resulting from nurturing that takes place in the family, and so on. Yet, we do not know how to measure or value much of what constitutes nonmarket production.

The Panel to Study the Design of Nonmarket Accounts was charged with evaluating current approaches, determining priorities for areas of coverage, examining data requirements, and suggesting further research to strengthen the knowledge base about nonmarket accounting.

SCOPE AND PRIORITIES

An overarching question for nonmarket account design is scope—where in the range of economic-related activities to draw the border of inclusion. The panel recommends the development of satellite accounts that cover productive inputs and valuable outputs that are not traded in markets, focusing on areas for which improved accounting would contribute to better policy making and to science. These accounts would provide a framework for examining difficult-to-measure activities that are excluded, or inadequately treated, in the NIPAs. Though one objective of nonmarket accounting is to support alternative aggregate measures of economic performance, satellite accounts are not intended to replace the current national accounts but to exist alongside them. Most of the work proposed by the panel would be conducted on an experimental basis and would not change the way the headline GDP is estimated.

While acknowledging that different users require different kinds of data and that new methods may be developed for valuing outcomes previously considered noneconomic, it is the panel's view that resources should initially be directed toward developing a more complete accounting of the population's productive activities rather than attempting to measure happiness or well-being. Throughout the report, the panel defends this position on practical measurement, as well as conceptual, grounds. Because improving output (and corresponding input) measures is a prerequisite to any vision for an expanded set of accounts, this is where the panel focused its energies.

The potentially valuable areas of nonmarket accounting are at different levels of development with respect to measurement concepts and available data. For that reason, the panel favors a staged approach. In general, the panel emphasizes areas for which new data sources offer opportunities to improve measurement of inputs or outputs, and excludes areas for which the likelihood of developing credible valuation estimates seems especially low. The staged approach allows work to proceed without commitment to a rigid framework, which might be difficult to agree upon across different areas of interest. Experimental methods—

potentially inconsistent with an integrated system—can be pursued in a framework of separate satellite accounts.

A number of productive activities are candidates for inclusion in a set of augmented accounts. Within the set of possibilities that might conceivably be of interest, the panel believes that priority should be given to the development of experimental accounts for areas that incrementally expand coverage of the conventional market accounts. This report focuses on five areas that the panel identified as being among the most promising and to which we would accordingly give high priority: household production, investments in formal education, investments in health, selected nonprofit- and government-sector activities, and environmental assets and services.

Household Production. The value of goods and services produced by households for their own consumption is quite likely large and, therefore, its measurement is essential for estimating a nation's overall level of economic activity. Evidence indicates that including household production as part of a nation's output also alters measured trend growth rates and their fluctuations over the business cycle. Given current knowledge and data, constructing an account for home production represents a logical early phase augmentation to the NIPAs. A major catalyst for this activity—and for prospects of future progress—is the development of time-use surveys, which generate data that are useful for estimating the extent and nature of work done in the home.

Investments in Formal Education and the Resulting Stock of Skill Capital. Although gross domestic product includes expenditures for education, it fails to adequately capture the contribution of related nonmarket activities to future economic growth, the well-being of individuals, and society in general. Because human capital, particularly that arising from education, is such a large component of the capital stock, a separate education account would contain essential data for improving research on investment, capital, and ultimately economic growth as measured by the traditional accounts. An education satellite account would incorporate market factors, as well as introduce experimental measures of nonmarket inputs and outputs. Successful development of an education account will require improved information on student and parent time inputs to schooling and further research on measuring and valuing the impact of education on the population.

Investments in Health and the Resulting Stock of Health Capital. A fully developed health account would enable researchers to estimate the effect that an array of inputs have on the stock of health and the value of associated changes in it. Additionally, measuring health is an important prerequisite for improving estimates of productivity in medical care. There are at least six major inputs into the production of health: medical care provided in market settings; care services provided without payment; time that individuals invest in their own health; consumption of other goods and services (some of which improve health and others of which are harmful) and nonmedical technology and safety devices; research and development; and environmental and "disease state" factors. There are two outputs of the health sector: the value of health capital, which can be defined as the expected flow of health consumption over the course of a person's remaining life, and the additional income that a healthier population generates.

Selected Activities of the Government Nonprofit Sectors. The initial focus of nonprofit-sector accounting should be on developing an account for tax-exempt organizations that are providers of public and charitable goods and services (as opposed to being providers of outputs to their members). The most quantitatively significant nonmarket input into nonprofit production is time, specifically that of volunteer labor. Therefore, construction of these accounts requires data on the number of volunteers and hours worked. The focus of experimental public sector accounting work should be on developing improved measures of output. In fact, full and independent (non-input based) valuation of goods and services is an important goal for comprehensive accounting for both the government and the economy's nonprofit institutions. Fulfilling this goal will entail more basic research.

Environmental Assets and Services. Environmental accounting has a long history. This panel is in agreement with the overarching recommendations of *Nature's Numbers* (National Research Council, 1999): Congress should authorize and fund recommencement of work with the ultimate goal of producing a comprehensive set of market and nonmarket environmental accounts. The accounts should focus on changes in values of the stocks of domestic natural resources and, probably more importantly in terms of nonmarket value, externalities associated with air and water pollution.

The areas listed above are substantial components of the economy—and the level of activity associated with each has the potential to change significantly over time—so focusing attention on them should improve our understanding of the nation's total production. Several of these areas overlap with coverage in the NIPAs and therefore may complement existing official statistics or help clarify policy issues based on those official accounts.

> **Recommendation 1.1:** The statistical agencies should develop a set of satellite accounts for household production, government and nonprofit organizations, education, health, and the environment. These accounts would provide a more complete picture of the nation's productive activities in these areas.

In addition to the five areas of nonmarket activity identified in this recommendation, the report includes a chapter on the role of the family in human capital development. Though something is known about the magnitude and the value of family inputs to human capital creation, the panel concluded that, given the current state of knowledge and data, it would be impossible to develop a comprehensive human capital investment account. Nonetheless, a fully specified set of nonmarket accounts would include a family care component or a more comprehensive human capital account, so this remains a worthwhile, albeit very long-term, goal. Such an account would formally recognize the investments families make in preparing children for lives as productive members of society, including the necessary foundation for later investments in formal education and health.

Though the panel considered other kinds of activities as well, we do not claim to have fully documented all areas of nonmarket production that contribute to social or private well-being. On the grounds hinted at above and elucidated further in the report, some other important areas—e.g., safety and security, leisure activities, and the underground economy—receive limited attention. This narrowing of scope notwithstanding, we believe that scholars and policy makers interested in a broad range of nonmarket topics will benefit from the principles laid out in this report.

Even within the set of areas identified here, there are differences in readiness to begin the construction of new satellite accounts. At this time, accounts for household production and the environment would rest on the firmest foundations; indeed, the Bureau of Economic Analysis and other national statistical offices already have done substantial work in these areas. In the remaining areas for which we advocate development efforts—the government and nonprofit sectors, education, and health—further conceptual thinking and new data collection are needed, and we encourage such work.

We acknowledge the difficulty of drawing boundaries between the areas of nonmarket activity we have identified as priorities. Improved health, for example, may result from better medical care, better education that contributes to better individual decisions about diet and exercise, or improved air and water quality. Nonetheless, the panel sees no realistic alternative to considering the different areas of nonmarket activity separately, but nonetheless recognizes the need to delineate the interactions and complementarities among them as the development of supplemental accounts proceeds.

Because the accounts proposed in this report will unavoidably overlap with one another (and with the market accounts), aggregate cost or product values will not be derivable by simply adding these accounts up—there would be extensive double counting. Nonetheless, given the current stage of development of frameworks and data relating to nonmarket activities, independent accounts will likely be more useful than a system that forces the accounting into a framework designed to eliminate overlap. Such a framework would require arbitrary allocation of costs or outputs across accounts.

CONCEPTUAL ISSUES

Complementing decisions about scope, a conceptual framework must be adopted on which to develop an economic account. For a number of reasons, the panel believes that experimental satellite accounts will be most useful if their structure is as consistent as possible with the NIPAs. Because the national accounts have undergone extensive scrutiny, reflecting a long history of research and policy use, the underlying principles are well tested and practice shows they can be implemented. Moreover, researchers are interested in developing augmented measures of output that are compatible with GDP. These considerations argue for pursuing an approach that uses dollar prices as the metric for relative value and, wherever possible, values inputs and outputs using analogous observable market transactions. Perhaps most importantly, nonmarket accounts should maintain a double-entry structure in which valuation of inputs to production (such as time) and outputs (such as child care) are based on separate price and quantity information.

> **Recommendation 1.3:** Nonmarket accounts should measure the value and quantity of outputs independently from the value and quantity of inputs whenever feasible.

Only with an independent measure of the value of output can we hope to address many of the questions for which nonmarket accounts could be most valuable.

A central issue is how to value inputs and outputs when market prices are absent. Quantitatively, people's time is the most important unmeasured input in nonmarket production.

> **Recommendation 1.4:** A replacement cost measure, adjusted for differences in skill and effort between nonmarket and market providers, should be adopted for valuing time inputs devoted to nonmarket production in cases in which someone else could have been hired to perform the work in question. Where this is not the case, an opportunity-cost-based measure, ideally adjusted to account for the intrinsic enjoyment associated with the activity in question, should be used for valuing time devoted to nonmarket production activities.

The report lays out in detail the contexts in which various input valuation approaches are appropriate.

As with the input side, valuation of nonmarket outputs should follow the principle of treating nonmarket goods and services as if they were produced and consumed in markets. Under this approach, the prices of nonmarket goods and services are imputed from a market counterpart. Even for the near-market cases, however, many goods exist for which the degree of replacement comparability—and, hence, substitutability—is not at all clear. More difficult yet are the cases for which a nonmarket good is an asset that has no direct market counterpart and is never sold. These kinds of outputs are prevalent in such areas as education and health and will require application of creative valuation methods.

DATA ISSUES

In addition to the many conceptual challenges that clutter the path toward construction of nonmarket satellite accounts, another barrier is the limited scope and quality of data available to support quantification and valuation of nonmarket activities in a way that is even remotely comparable to that which is done for the NIPAs. Fortunately, there is recent progress to report.

The new American Time Use Survey (ATUS) will provide rich information about the most important input to nonmarket production—people's time.

> **Recommendation 2.1:** The American Time Use Survey, which can be used to quantify time inputs into productive nonmarket activity, should underpin the construction of supplemental national accounts for the United States. To serve effectively in this role, the survey should be ongoing and conducted in a methodologically consistent manner over time.

What makes the ATUS particularly valuable for the purpose of creating nonmarket accounts is that its information will be provided year after year.

Despite the tremendous step forward represented by the ATUS and its role as an essential building block for most nonmarket accounts for the United States, there are concerns about its reliability and validity for this purpose. Lower-than-expected response rates—and the possible resulting nonresponse bias—are worrisome. Efforts to assess the extent of biases in survey responses—and, if necessary, to confront that bias by raising response rates or making appropriate adjustments to the estimates—should be a priority.

> **Recommendation 2.2:** The Bureau of Labor Statistics should commit resources adequate to improve response rates in the American Time Use Survey and to investigate the effects of lower-than-desirable response rates on survey estimates.

Demographic data go hand in hand with time-use data in laying the foundation for nonmarket accounts: time-use data can be used to answer questions about what individuals with particular characteristics are doing with their time; demographic data describe the distribution of characteristics in the population. Although we recognize that creation of better coordinated demographic data would require significant effort by statistical agencies and other suppliers of data, the development of such data is an important goal.

> **Recommendation 2.4:** A consistent and regularly updated demographic database should be assembled as an input to nonmarket accounts. The database should include information on the population by age, sex, school enrollment, years of education and degrees completed, occupation, household structure, immigrant status, employment status and, possibly, other dimensions.

The fact that high-quality demographic data are already collected by the Census Bureau and other agencies improves the feasibility of implementing this recommendation. The American Community Survey also will help advance the effort to produce a more fluid demographic description of the population. Given budget constraints of the statistical agencies, the demographics database should be built, to the maximum extent possible, using existing data.

Full development of nonmarket accounts also requires further research and data development directed toward defining and measuring output. It is not always obvious what the outputs associated with various nonmarket activities are, especially in such service-oriented areas as education, health, social services, culture and the arts, and recreation. Frequently, in difficult-to-measure sectors, the value of output is not measured directly, but set equal to the aggregate value of the inputs used in its production. Accordingly, little is known about growth, quality improvements, or productivity in these sectors. The data that will be needed to create independent output measures are discussed throughout the report. We offer numerous recommendations on this topic and more generally about the next steps needed in research and data development to advance nonmarket accounting.

1

Introduction

BACKGROUND AND OVERVIEW

Gross domestic product is the nation's most influential measure of overall economic activity and growth. The national income and product accounts (NIPAs) that underlie gross domestic product, together with such other key economic data as price and employment statistics, are widely used as indicators of how well the nation is doing.

Since their earliest construction for the United States by Simon Kuznets in the 1930s (Kuznets, 1934), concerns have been voiced that the national income and product accounts are incomplete. The venerable NIPAs meet rigorous standards and enjoy broad acceptance among data users interested in tracking economic activity. These accounts are, however, primarily market-based and, by design, shed little light on production in the home or in other nonmarket situations. Furthermore, even where activity is organized in markets, important aspects of that activity may be omitted from the NIPAs. In some cases, unpaid time inputs and associated outputs are integral to production processes but, because no market transaction is associated with their provision, they are not reflected in the accounts. One illustration is provided by estimates (LaPlante et al., 2002) suggesting that the value of in-home long-term care services provided by family and friends is greater than the value of similar market-provided services.

In other areas, the output resulting from market-based production may be incorrectly characterized or valued. There is wide agreement, for example, that the output of the education sector properly should be considered investment rather than consumption, and that its value should be assessed in terms of the returns on that investment rather than the cost of the inputs used in its production. The conventional accounts do not include changes in the asset value of human capital production associated with education, health care, and other personal investment activities. Available estimates are rough, but suggest that the value of the human capital stock may be as large as that of the physical capital stock (see Kendrick, 1967, and, for a more recent discussion in the context of analyzing economic growth, Mankiw et al., 1992).

Although the importance of nonmarket—but productive—endeavors has long been recognized, few attempts have been made to provide systematic information about even the most quantitatively significant of them. Economic accounting need not, and should not, extend to all nonmarket activities, but there are certain areas in which nonmarket accounts, designed to supplement the NIPAs, could make particularly important contributions. We stress the potential value of new methods of accounting for home and volunteer production efforts, education, health, and environmental improvement or pollution.

Given that limited coverage of the NIPAs has long been recognized, why is this report needed at this time? The existence of economically valued nonmarket inputs and outputs is already widely recognized. Examples are easy to find: the higher value of a house sold after improvements are made by the homeowner; the meals that a nonprofit soup kitchen serves to the homeless; or the increase in the productive capacity of the economy attributable to extensions of working life resulting from modern diabetes treatments. But information about these productive activities is not systematically compiled and routinely updated as are the NIPAs and other market-based accounts. The state of nonmarket accounts today resembles the situation for market-based accounting in the 1920s and 1930s before the creation of the NIPAs—new data were becoming available, but they were not organized and published in a systematic accounting framework. Advances in data collection and in economic analysis of nonmarket activities merit the review that appears in this volume. One event alone—the development and recent publication of the American Time Use Survey—justifies a new round of thinking about nonmarket accounting issues. In this report we hope to encourage social scientists to pursue the analysis of nonmarket activities and the development of corresponding data collection and accounting systems. We also point out new ideas and new data sources that have improved the prospects for progress.

The Panel to Study the Design of Nonmarket Accounts was charged with evaluating current approaches, determining priorities for areas of coverage, examining data requirements, and suggesting further research on nonmarket accounting. The panel's charge includes four specific tasks:

- to review efforts to develop nonmarket accounts by government agencies, as well as by private organizations and scholars, including theoretical as

well as empirical studies and actual implementation of nonmarket accounting frameworks;
- to make specific recommendations on the framework and sectors for developing nonmarket accounts and determine a set of priorities with respect to developing or phasing nonmarket accounts;
- to examine and make recommendations with respect to key data that are needed to develop nonmarket accounts, considering such efforts as the Bureau of Labor Statistics's (BLS) time-use survey; and
- to investigate and make recommendations for research related to nonmarket accounts in the areas of statistics, economics, psychology, survey research, and other disciplines.

The panel worked over a period of $2^1/_2$ years to fulfill this charge. Our recommendations aim to provide practical guidance for constructing nonmarket accounts. Over the course of its deliberations, the panel found that the conceptually ideal approach often does not align with what can realistically be expected in practice. The panel has tried to be clear about when its recommendations reflect compromise required by practical constraints, such as data availability or measurement feasibility.

The rest of this chapter discusses the motivation for this study, priorities for nonmarket accounting, and issues relating to account scope and conceptual framework. Chapter 2 identifies and describes accounting and data issues—particularly issues related to the measurement of time use—that are key to both market and nonmarket accounts. Chapters 3 through 8 provide detailed analyses of the individual areas for which the panel has evaluated development of nonmarket satellite accounts.

The chapters on specific satellite accounts each begin with a discussion of the area, considering its importance and the role that nonmarket activity plays in it. As noted above, in many of the areas the panel has chosen to explore, market activity coexists with nonmarket activity. Students pay tuition to attend universities (a market transaction), but there is no market transaction that captures directly the value of the time they devote to this endeavor. Likewise, nonprofit organizations may hire employees (a market transaction), but also may use unpaid volunteers, the value of whose time and output produced is nowhere reflected.

Central to each of the chapters is a discussion of the conceptual issues related to the measurement of nonmarket activity in the area it covers. These issues lead to consideration of how inputs—both market and nonmarket—and resulting nonmarket outputs might be quantified. The transformation of inputs into outputs presupposes a production technology and, in a number of the chapters, this technology is also a subject of discussion.

Another matter that figures prominently throughout the report is how to attach values to specified quantities of nonmarket inputs and outputs. The panel's approach, in essence, is to seek prices that can be assigned to the identified quantities. In some cases, satisfactory price measures can be found; in other cases, available information may be unsatisfactory, either because it is conceptually deficient or because it is insufficiently precise to be useful in the implementation of an accounting scheme. Each chapter devotes substantial attention to the question of whether and how satisfactory price measures might be developed. In many cases, researchers have already developed input or output valuations of the sort the panel contemplates.

On the basis of these topic-by-topic evaluations, the panel makes recommendations regarding areas to be viewed as high priorities for the development of nonmarket accounts and areas to be viewed as lower priorities. Finally, for those areas in which the panel believes further work holds promise, the individual chapters include recommendations about ideal data for the intended purpose and steps that might be taken toward the production of such ideal—or at least better—data for the United States.

MOTIVATION

Extending the nation's accounting systems to better incorporate nonmarket production promises substantial benefits to policy makers and researchers. One objective of improved nonmarket accounting is to support alternative aggregate measures of economic performance. Nonmarket accounts would enhance the ability of researchers (and statistical agencies) to produce augmented gross domestic product (GDP) measures that reflect a broader array of outputs. Data from nonmarket accounts could provide the information required to construct appropriate price deflators needed for augmented real GDP calculations and productivity measurement.

The inherent limitation of the NIPAs—that they fail to consider the full array of the economy's productive inputs and valued outputs—might not be especially important if market and nonmarket activities trended similarly, but this cannot be assumed. To take one often-cited example, failing to account for the output produced within households may lead to misleading comparisons of economy-wide production, as conventionally measured. According to BLS figures, the female labor force participation rate in the United States has grown substantially, from about 34 percent in 1950 to almost 60 percent in 2000. To the extent that the entry of women into paid employment has reduced effort devoted to household production, the long-term trend in output as measured by GDP may exaggerate the true growth in national output (Landefeld and McCulla, 2000). Similarly, the relatively smaller portion of total output attributable to home production in the United States as compared to many developing countries may exaggerate its national output relative to theirs.

Perhaps less well recognized are potential problems with the measurement of national output over the business cycle. If people who lose their jobs during cyclical downturns take advantage of their absence from paid employment to increase the effort they devote to home production, the short-term decline in national output may be dampened relative to that measured by GDP (Greenwood et al., 1995). Knowing more about the level and distribution of nonmarket activity could be important for other purposes. Such information could, for example, change perceptions of the extent of economic inequality among U.S. households and how that has changed over time. This, in turn, could affect where welfare and poverty lines are drawn (Michael, 1996).

Accounting for nonmarket activity also would contribute to our understanding of the sources of economic growth. Researchers studying this topic have long found it necessary to supplement data from the national accounts with external estimates of the contributions of research and development, investments in human capital, and the services of the natural environment (Mankiw et al., 1992). Economists and historians have shown that, over the last few centuries, factors such as technical change, scientific inventions, and discoveries in medicine—many of which have a nonmarket dimension—have accounted for a very large portion of the growth in living standards (Fogel, 1986; Nordhaus, 2003; Cutler, 2004). Historical trends reveal the reality that neither economic production nor contributions to social welfare take place exclusively within the market's border, but extend to many nonmarket activities.

Nonmarket accounting also would illuminate the processes whereby inputs are transformed into outputs in particular sectors. Consider, for example the production of health. Currently constructed health *expenditure* accounts track market payments but do not identify the outputs in a way that is very useful for measuring price change or productivity. In contrast, a health account would relate health improvements—the real "good" that is produced—to medical treatments, as well as to a wide range of other inputs, including diet, the environment, exercise, and research and development. By most measures, improvements in health have outpaced the increase in spending on medical care. Since medical care interacts with this broad range of interrelated factors, however, we do not know with any certainty the productivity of resources directed into health care (Cutler and Richardson, 1997; Cutler, 2004). Optimally, expenditures and outcomes would be tracked so that changes in people's health could be linked to different actions; in turn, this information could support better management of expenditures (both private and public) to achieve desired outcomes.

To take another example, education accounts might be designed to relate improvements in skill capital—the output—to the various inputs of the educational process. As in the health case, schooling is characterized by a mix of market and nonmarket inputs and outputs. In education, the value of time students spend in school is likely to be of the same order of magnitude as expenditures on marketed inputs. The *2003 Statistical Abstract* shows that, in 2000, school expenditures on primary and secondary education amounted to approximately $400 billion and that just over 47 million students were enrolled in primary and secondary schools. Assuming 180 days at 6 hours a day, plus 1 hour of commuting time and 2 hours of homework per student, students in these grades devoted more than 75 billion hours to their education. If students' time were valued at $5.35 per hour (purely for illustrative purposes), the value of unpaid student time would be approximately as large as the expenditures measured in the conventional accounts.

This report identifies and discusses many of the key and sometimes controversial issues, such as those hinted at above, that underlie nonmarket accounting. One goal of the report is simply to remind readers of the major omissions built into our system of economic measurement. In so doing, the panel hopes to encourage contributions by social scientists to improve the measurement of nonmarket activity and to point out new ideas and new data sources that have improved the prospects for progress. Time is the dominant input to nonmarket production, and the lack of good measures of how people spend their time has seriously handicapped work in this area. The panel is optimistic that the newly developed American Time Use Survey, produced by the BLS, will spur new work to develop informative nonmarket accounts.

PRIORITIES, SCOPE, AND OBJECTIVES

Scope of Coverage in the NIPAs

Traditional national income and product accounts include primarily the output of marketed goods and services—that is, goods and services that are bought and sold in market transactions. An international standard, the System of National Accounts (SNA) (Commission of the European Communities et al., 1994), provides a set of guidelines, accounting rules, classifications, and definitions designed to, among other things, facilitate international comparison of market-based measures. The boundaries to national economic accounting employed in the U.S. NIPAs largely conform to the standards outlined by the SNA.[1]

At the time of their development, there were good reasons for the U.S. national income and product accounts to focus on market production. In assessing the disruptions attributable to the World Wars and the Great Depression, policy makers understandably had a primary interest in how these events had affected market activity. In addition, as recognized by Kuznets and others, reliable measurement of nonmarket production would have been extremely difficult. Nonetheless, there have long been influential economists who argued that economic accounting should extend beyond the market. One of the great founders of modern economics, A.C. Pigou, wrote that national accounts should include elements that reflect economic welfare and that can "be brought directly or indi-

---

[1]Additional background information on the structure and scope of the NIPAs is provided in Chapter 2.

rectly into relation with the measuring rod of money" (Pigou, 1920, p.11). He emphasized that the word "can" might mean anything from "can easily" to "can with mild straining" to "can with violent straining." National accounting practices in most countries lean far more toward those elements that "can easily" be measured in money terms than those that can be measured only with "violent straining."

Though the national accounts produced by the Bureau of Economic Analysis (BEA) generally exclude activities that do not involve a market transaction or produce a marketed output, there are exceptions—most notably the imputation for the rental value of owner-occupied housing. This imputation is based on assumptions that are approximately as crude as those for, say, valuing the time spent cleaning a house at the price a cleaning service would charge. One reason for making an imputation for the value of owner-occupied housing is to ensure that the accounts are invariant to trends in home ownership (which, incidentally, has increased significantly in the past half-century; see U.S. Census Bureau, 2004). Other imputations for nonpriced, nonmarketed items in the NIPAs include those for wages and salaries paid in kind, food and fuel consumed on farms, and the services provided by banks, insurance companies, and other financial intermediaries that are not reflected in explicit service charges. The imputations for banking services are somewhat unique. In banking, there are observable market transactions that provide an estimate of the nominal value of banking output. Imputations are necessary, however, to allocate the nominal value of unpriced services between borrowers and depositors (see Fixler et al., 2003, for a more complete discussion). We are not aware of any body of analysis that would set the outer bounds of market accounting at precisely the point chosen by the SNA or by the BEA.

One key characteristic of the nonmarket items that are covered in conventional accounting systems is that their consumption is closely related to the sales and purchases of marketed goods and services, making estimation reasonably straightforward if not always precise. For example, the rental value of owner-occupied housing is imputed from observed rents for similar housing. Similar, straightforward imputations could serve as a basis for price estimation for some nonmarket items excluded from coverage in the national accounts but, for many others, such a close comparison will not be possible.

While the national accounts exclude the output resulting from many areas of nonmarket activity, information relevant to these activities often is included. In many cases, purchases of inputs that contribute to nonmarket outputs are treated as expenditures for final demand. Spending on food, cleaning supplies, and laundry detergent is counted as part of personal consumption. In an accounting system that considered meals, house cleaning, and laundry services as elements of consumption, these kinds of expenditures would appear as intermediate inputs to the production of consumption goods. Similarly, government expenditures on education are included in the accounts. These would become intermediate products if the human capital produced through educational activities were treated as the relevant output. And many of the inputs to medical care are included, but the accounts contain relatively little separate information about the value of the services provided or the health capital formed.

For the cases we have elected to explore, the conventional accounts generally do not reflect the full range of inputs used in the production of the output of interest. And in no case is the value of the resulting output, whether goods and services produced for current consumption or the creation of a productive asset, measured fully and independently of the value of the inputs used in its production. Independent measurement of output entails estimating prices and quantities of that output, as opposed to building up valuations from cost side factors.

Satellite Accounts

While acknowledging the limitations in their coverage, the panel is not proposing that the scope of the existing NIPAs be expanded to encompass currently omitted nonmarket activities. The existing core accounts have the important virtues of consistency over time, hard-won comparability across countries, and, aside from a limited number of agreed-upon imputations deemed necessary for certain nonmarket activities, solid grounding in observed market transactions; we would argue strongly that they should be preserved. Likewise, the statistical agencies should not develop a comprehensive measure of total economic output as a substitute for the existing GDP measure. Even at a conceptual level, there is no consensus about what an appropriate replacement for the existing measure of GDP might be. We expect that data from the satellite accounts we advocate will allow researchers to develop a variety of expanded GDP measures. It would be premature, however, to endorse a specific measure—indeed, it seems unlikely that a consensus around any one measure ever will develop.

Instead, we recommend the development of satellite accounts to report on selected activities not included in the conventional accounts. Satellite accounts can link to the existing economic accounts as appropriate, but also expand into areas that the NIPAs do not cover. Furthermore, satellite accounts can be developed even where standards of accuracy and data quality are not up to the level of the NIPAs, without compromising the conceptual basis or technical integrity of the conventional accounts. Similarly, where no consensus yet exists regarding the best way to measure a particular area of nonmarket activity, satellite accounts allow for the flexibility to experiment with alternative methodologies that might go against convention. The goal is to extend the accounting of the nation's productive inputs and outputs, thereby providing a framework for examining the production functions of some difficult-to-measure nonmarket activities.

The idea of satellite accounts is not a new one. The BEA has long conducted research on topics beyond the scope of the conventional accounts. A representative BEA definition of satellite accounts is as follows (Bureau of Economic Analysis, 1994a, p. 41):

> [S]atellite accounts are frameworks designed to expand the analytical capacity of the economic accounts without overburdening them with detail or interfering with their general purpose orientation. Satellite accounts, which are meant to supplement, rather than replace, the existing accounts, organize information in an internally consistent way that suits the particular analytical focus at hand, while maintaining links to the existing accounts. In their most flexible application, they may use definitions and classifications that differ from those in the existing accounts. . . . In addition, satellite accounts typically add detail or other information, including nonmonetary information, about a particular aspect of the economy.

The System of National Accounts offers a similar description (Commission of the European Communities et al., 1994, pp. 45, 489):

> Satellite accounts provide a framework linked to the central accounts and which enables attention to be focused on a certain field or aspect of economic and social life in the context of national accounts; common examples are satellite accounts for the environment, or tourism, or unpaid household work. . . .Satellite accounts or systems generally stress the need to expand the analytical capacity of national accounting for selected areas of social concern in a flexible manner, without overburdening or disrupting the central system.

The panel views the accounting frameworks described in this report as harmonious with these definitions and concludes that, for a number of industries and sectors with significant nonmarket components, satellite accounts can be developed that will generate meaningful and useful data to inform policy and to advance research.

Measurement Objectives

Although they are widely used in comparisons intended to shed light on trends and relative levels of national well being, the national accounts are not designed to measure the welfare of the population or the contributions to welfare associated with particular sectors. Rather, the NIPAs provide a measure of the nation's market output (and income). Output obviously correlates to some extent with consumers' welfare—indeed, there is a well-developed literature in cost-benefit analysis showing that, under certain conditions, changes in the former may approximate changes in the latter[2]—but national income is not the same as national welfare.

---

[2]Dreze and Stern (1987) show that, under certain conditions, the net social benefits of a "project" (that is, some change in output) are equal to the change in GDP measured at market prices.

For some potential policy or research applications, an account based on a traditional (price times quantity) valuation may be most useful; for others, an account oriented toward "social welfare valuation" might be conceptually preferable. If resources were unlimited, multiple accounts would be developed to serve all of the various purposes that arise. But this is not the case, and development of separate accounts to meet all possible needs is not a realistic option.

Certain nonmarket activities—much household work, for example—result in the production of goods and services very similar to those sold in markets. These goods and services presumably are welfare enhancing as well. A logical first step in developing a set of satellite accounts is to quantify and value these "near-market" activities. If coverage of the nonmarket accounts were limited to items for which there are close market substitutes, however, many factors that affect well-being would be excluded.

Broader accounting approaches also can be envisioned. One option would be to include nonmarket activity if it is possible to conceive of a procedure that would reveal the monetary value that people place on it. This approach would lead to a more expansive set of nonmarket accounts. Under this definition, an activity such as leisure might be considered in scope. Even such a seemingly noneconomic outcome as "family disintegration" could enter the accounting structure if a method could be developed that revealed what society would be willing to pay to prevent its occurrence. Our ability to conceive of such procedures has increased in the years since Becker's seminal work on marriage and the family (e.g., Becker 1973, 1974), and the boundary between what is and is not a proper candidate for inclusion, as determined by the rule that an entry must receive a monetary valuation, undoubtedly will change over time.

As noted above, one of the important potential uses of satellite accounts for nonmarket activities is to support broader aggregate measures of national output. A leading example of an alternative aggregate output measure is James Tobin and William Nordhaus's measure of economic welfare (Nordhaus and Tobin, 1972). Although various alternatives exist, there is no professional consensus around any single expanded measure, and we do not believe that such a measurement objective provides the organizational principles for a system of satellite accounts. A more realistic approach to developing augmented measures of output would begin with a set of largely independent accounts covering a wide range of nonmarket activities.

While acknowledging that different users require different kinds of data and that new methods may be developed for valuing outcomes previously considered noneconomic, it is the panel's view that resources should initially be directed toward developing a more complete accounting of output rather than attempting to measure happiness or well-being. This position can be defended on practical measurement, as well as conceptual, grounds. Accounting for nonmarket activities and results that parallel those already represented in the national accounts would enable the developers of the new accounts to build more directly on past experience. Such things as household production of cleaning services or participation in educational activities have market analogues and are (relative to, say, leisure) more closely aligned with what is currently viewed as output in the accounts, meaning that the hurdles to clear in devising sensible measurements are lower. Because improving output (and corresponding input) measures is a prerequisite to any vision for an expanded set of accounts, this is where the panel focused its energies. Even if the accounting objective is limited primarily to measurement of output, there is still a formidable amount of work to be done.

Nonmarket Accounting Priorities

In defining the boundary of what should be counted when developing measures of nonmarket output, some have advocated the application of Margaret Reid's (1934) third-party criterion: Is the output something that a person could have hired someone else to produce? This criterion seems appropriate for certain areas—for example, a household production account could be designed to include such things as meals, clothing services, shelter services, and a basic component of child care, but to exclude fertility, studying, and exercise. For other areas of nonmarket activity, such as education and health, the third-party criterion is clearly inappropriate: there is no replacement for self involvement in the activities required to enhance a person's cognitive skills or improve his or her health—activities that nonetheless produce valuable but nonmarketable capital outputs.

In considering how broadly nonmarket accounting work should be cast, one must ask what standards of accuracy and reliability should be applied to measures of inputs and outputs. We do not have a full answer to this question. Traditionally, the statistical agencies responsible for economic accounting—the Bureau of Economic Analysis of the Department of Commerce and the Bureau of Labor Statistics of the Department of Labor—set high standards of accuracy. At least initially, nonmarket satellite accounts presumably will have to be constructed under a more forgiving standard, but there undoubtedly will be debate about how imprecise a number can be and still remain useful. In market accounting, outside researchers often produce accounts from ingredients supplied by the statistical agencies—for example, a consulting firm publishes monthly GDP from data supplied by the BEA, though the BEA does not consider the data to be sufficiently reliable in all sectors to produce an official monthly measure. We anticipate similar private accounting efforts with nonmarket data.

As detailed in this volume, potentially valuable areas of nonmarket accounting are at different levels of development with respect to measurement concepts and available data. For that reason, the panel favors a staged approach. Work should begin in areas where potential gains are high and costs (mainly in terms of new data collection needs) are low. For example, we believe that accounting for certain types of home production should begin soon, based on data from the American Time Use Survey. In contrast, neither the conceptual understanding nor the necessary data are available to support an accounting of the human capital formed at home when parents teach their toddlers to talk or to think about the effects of their actions on others. The staged approach allows work to proceed without commitment to a rigid framework, which might be difficult to agree upon across different areas of interest. Experimental methods—potentially inconsistent with an integrated system—can be pursued in a framework of separate satellite accounts.

A wide range of productive activities seems worthy of exploration for possible inclusion in a set of augmented accounts that would reflect more fully and more directly the breadth of the economy's productive activities. Within the set of outputs that might conceivably be of interest, the panel believes that priority should be given to the development of experimental accounts for areas that incrementally expand coverage of the conventional market accounts. The chapters that follow discuss several specific areas that the panel believes to be among the most promising and to which we accordingly give top priority:

- household production,
- investments in formal education and the resulting stock of skill capital,
- investments in health and the resulting stock of health capital,
- selected activities of the government and nonprofit sectors, and
- environmental assets and services.

The panel believes that improved information about these selected areas of nonmarket activity would be of significant value to both policy makers and researchers. Moreover, given the conceptual tools and data that exist already, several of these areas are ripe for development.

Recommendation 1.1: The statistical agencies should develop a set of satellite accounts for household production, government and nonprofit organizations, education, health, and the environment. These accounts would provide a more complete picture of the nation's productive activities in these areas.

The areas listed above are substantial components of the economy—and the level of activity associated with each has the potential to change significantly over time—so focusing attention on them should improve our understanding of the nation's total production. As discussed in more depth in the chapters that follow, all are areas of likely interest in their own right to a range of data users. Several of the areas overlap the NIPAs and thus complement existing official statistics.

The panel's focus on household production, government and nonprofit organizations, education, health, and the environment also reflects a feasibility constraint. Though measurement will be far from easy or noncontroversial for almost any nonmarket satellite account that could be envisioned, the set chosen for study excludes areas for which sensible approaches to quantifying and valuing inputs or outputs appear especially far from reach. We also have given priority to areas for which emerging data sources offer new opportunities. For example, there is considerable appeal to the idea of a fully developed human capital account that would parallel the extensive efforts by statistical agencies to measure the value of the nation's physical capital stock. Given the data and conceptual work that is needed before real progress toward this ideal can be made, however, the panel advocates focusing first on a piece of the puzzle, a formal education account, where research and data on time inputs and output valuation are further developed.

Even within the set of areas identified here, there are differences in readiness to begin the construction of new satellite accounts. At this time, accounts for household production and the environment would rest on the firmest foundations; indeed, both the BEA and other national statistical offices already have done substantial work in these areas. In the remaining areas for which we advocate development efforts—the government and nonprofit sectors, education, and health—further conceptual thinking and new data collection are needed, and we encourage such work.

A set of accounts that covered the areas described in this report would go a long way toward documenting nonmarket production that contributes to social or private well-being, and would address many of the key principles generalizable to nonmarket accounting broadly. Nonetheless, we make no claim that a set of accounts covering the five areas on which the panel recommends focusing work would measure all nonmarket production or fully document all of the assets that contribute to social or private well-being. On the grounds outlined above (and elucidated further below), some other important areas receive limited attention in this report. They include safety and security, leisure activities, and the underground economy. We also do not delve deeply into the burgeoning literature that has grown up around ideas associated with the term "social capital." Social capital has been loosely defined as "the networks of social relations that provide access to needed resources and supports. . . . A person's family, friends, and associates are part of social capital. . . . Determinants of social capital formation can operate both at the individual and at the community level" (Policy Research Initiative, 2003).[3]

Consistent with this view that the satellite accounts should expand incrementally from the conventional accounts, the value of pleasure derived from the process of producing nonmarket outputs also is not part of the nonmarket satellite accounts the panel proposes. For example, the pleasure a person derives from cooking at home should not be part of a household production account, just as the pleasure from working as a professional chef is not included in the NIPAs. Even if one did want to account in some fashion for the pleasure associated with

[3]See the useful critical review by Steve Durlauf (2004). Many of the ideas associated with social capital were popularized in Robert Putnam's *Bowling Alone*.

nonmarket activity—and a satisfactory method of quantifying its value could be derived—we believe the results would best be recorded in a separate account, rather than intermingled with other dissimilar outputs in an account focused on, for example, home production of goods and services, education, or health.

Related questions exist concerning what constitutes an input to nonmarket production. In particular, how should the time devoted to consumption be treated? Enjoying a restaurant meal, for example, requires not only the meal itself, but also the time of the diner who consumes it. Should that time be counted as an input to nonmarket production? Again, our view is that valuing time spent in consumption should not be a priority for near-term nonmarket accounting efforts.

We acknowledge that it is difficult to draw boundaries between the areas of nonmarket activity we have identified as priorities. Improved health, for example, may result from better medical care, better education that contributes to better individual decisions about diet and exercise, or improved air and water quality. Identifying the full set of inputs to improved health outcomes thus may be difficult, and some of these inputs also may contribute to other desirable outputs. To take another example, additions to the stock of human capital may flow not only from investment that occurs within the formal educational sector, but also from investments that occur within the home; thus, they might be considered a form of home production. The panel sees no realistic alternative to considering the different areas of nonmarket activity separately, but nonetheless recognizes the need to delineate the interactions and complementarities among them as the development of supplemental accounts proceeds.

Because the accounts proposed by the panel in this report will unavoidably overlap with one another (and with the market accounts), aggregate cost or product values will not be derivable by simply adding these accounts up—there would be extensive double counting. Nonetheless, given the current stage of development of frameworks and data relating to nonmarket activities, independent accounts will likely be more useful than a system that forces the accounting into a unified framework designed to eliminate overlap. Such a framework would require arbitrary allocation of costs or outputs across accounts. Furthermore, the new accounts are likely to proceed at different speeds of development. As a result, the amount of double counting and interactions will depend on which accounts are developed or in use at a point in time, raising the specter that the exercise would have to be redone each time a new account is developed.

A final question concerns the periodicity for which nonmarket accounting data should be compiled. On one hand, part of the value of any satellite account lies with its contribution to assessing economic trends and forecasting the impact of alternative policy options. This requires that updated information based on a consistent methodology be provided on a regular schedule. On the other hand, given the limitations of available data, it seems unlikely that quarterly (much less monthly) accounting generally would be feasible and meaningful. The appropriate publication schedule may be annual in some cases and perhaps even less frequent in others, but the panel believes it is important that there be a regular schedule.

> **Recommendation 1.2:** In parallel with the NIPAs, the satellite nonmarket accounts should be based on a consistent methodology over time and be published on a regular schedule to produce a consistent time series of comparable data.

A CONCEPTUAL FRAMEWORK

The national accounts have proven extraordinarily useful as a vehicle for monitoring and studying the evolution of the economy. They have the intentional restriction, of course, that they do not systematically incorporate nonmarket activity, but their strong conceptual foundations provide an example that any newly developed nonmarket satellite account would do well to take as a starting point. Because the national accounts have undergone such extensive scrutiny, reflecting research and policy use over many decades, the underlying principles are well tested and practice shows they can be implemented. Many of the methodological issues that will arise in designing satellite accounts have analogs in the national accounts (Nordhaus, 2004, p. 3). Moreover, the heavy reliance of policy makers and others on the existing accounts, together with the interest researchers have in developing augmented measures of output that are compatible with GDP, imply that any supplemental accounts will be most useful if the information they contain is as consistent as possible with the information in the NIPAs.

What specifically does this imply? The NIPAs rest on a double-entry structure that values outputs independently of inputs and incorporates measures of quantity and price for both. One of the most important applications of the national accounts is the measurement of productivity growth, which requires these separate measures. The NIPAs use dollar prices as the metric for relative value; value outputs at their marginal rather than their total value; and derive these marginal values wherever possible from observable market transactions. Following comparable practices in the nonmarket accounts would facilitate comparisons between those accounts and the NIPAs.

The national accounts report three measures for each type of product at the most detailed level: the quantity, the price, and the dollar value. These are linked by the principle that value is price multiplied by quantity. With few exceptions, the accounts obtain data on value from primary sources, and quantity is calculated by dividing value by a measure of price. In a few cases, data on value and quantity are obtained, and price is calculated as the ratio of the two. We anticipate that similar calculations would be used in satellite accounts. In addition, a satellite account might use data on quantity together with estimates of prices to calculate value as the product of the two. This procedure seldom is necessary in the national accounts because value data are generally available from primary sources.

Implications of Double-Entry Bookkeeping

One of the strengths of the NIPAs is the double-entry bookkeeping used in their construction. Independent estimates of total output are developed on the basis of the dollar value of output sales, on one hand, and the dollar value of payments to factors of production, on the other. In principle, these two independently derived sums—the product side and the income side estimates of GDP—should be equal. The difference between the two estimates is the statistical discrepancy, which by construction differs from zero only because of measurement errors. In the conventional accounts, a small statistical discrepancy suggests that the value of output has been well measured, since two independent measurement methods give approximately the same answer; a larger statistical discrepancy signals the existence of measurement problems.

Interpretation of the difference between input costs and output values is somewhat less straightforward in the case of a nonmarket account. In a competitive market context, an inefficient firm—one for which the value of the resources employed exceeds the value of the output produced—will eventually be driven out of business. Competitive pressures do not operate in the same way in the nonmarket context. That households optimize with respect to their allocation of time is a more tenable assumption than the alternatives, but households that fail to optimize are not driven out of business and may continue to exist indefinitely. This introduces the possibility that, depending on how it is measured, the cost of time devoted to home production could exceed or fall short of its productive value.

The conceptual equality of output values and input costs in the market accounts also reflects the convention that is employed for measuring capital costs. Revenues not spent on other costs of production are considered to be a part of the cost of capital; put differently, capital is treated as the residual claimant. An alternative approach to valuing capital services—and one that seems applicable to the nonmarket accounts—would be to use a standard measure of the flow cost of capital. Such a measure would be constructed using information on the value of the capital employed to produce the nonmarket output and an assumed market rate of return to that capital. Using this approach, the cost assigned to capital services could be greater or less than their productive value.

Capital-market constraints, such as those that might arise from lenders' reluctance to finance the production of assets that cannot be marketed and therefore cannot readily serve as loan collateral, may be particualrly important in the nonmarket context. Absent capital market constraints, larger investments might be made. Because the amount of investment is constrained, however, the return on investments that do occur will be higher than the market rate of return. Valuing nonmarket investments in a fashion that ignores this possibility by, in effect, imposing an assumed normal rate of return—for example, valuing educational output based on the costs of inputs employed—could lead to a figure less than the true value of the asset produced.

Differences in technology or scale of production between nonmarket and market production are other possible reasons for divergence between the cost of inputs and the value of output in nonmarket production. It might be more efficient, for example, to prepare ten meals rather than to prepare one; unless they belong to a large family, however, individuals cooking at home cannot take advantage of this scale economy, and reasonable estimates of the value of resources used to prepare the meal at home might exceed the market value of a restaurant meal. The transactions costs associated with traveling to dine at a restaurant, however, might still make it more attractive to cook and eat at home.

Though the sum of the values of the inputs used to produce a nonmarket output may provide a poor estimate of the value of that output, this has commonly been the practice in some measurement areas. It is, for example, by far the most common approach in the literature on valuing government services and home production (see National Research Council, 1998; Holloway et al., 2002). Well-designed input-based output valuations are a clear improvement over ignoring nonmarket activity altogether. Only with an independent measure of the value of nonmarket output, however, can one hope to address many of the questions for which nonmarket accounts could be most valuable.

> **Recommendation 1.3:** Nonmarket accounts should measure the value and quantity of outputs independently from the value and quantity of inputs whenever feasible.

For the reasons given, the panel advocates the adoption of the double-entry bookkeeping of the NIPAs for use in any satellite accounts, even if it is not operationalized in exactly the same way in the nonmarket context. For some areas—particularly health, for which output measurement is especially difficult—input and output measurement will not develop in tandem. This should not be a deterrent to accounting efforts in these areas—a one-sided account is generally better than no account at all. For example, an input-based account for formal education based on imputed values of student time would be useful even if it did not measure the value of the output of education independently. Similarly, an accounting of volunteer labor in the economy could provide useful data for research and policy.

Classifying Goods and Services

Several efforts to modify or otherwise expand the national accounts have originated from the belief that misclassifications in the present accounts give a false impression of economic activity. For example, one could argue that at least some governmental activities (e.g., protection and inspection services) properly should be treated as inputs to business activity rather than as an output of the economy, as is current practice.[4] Similarly, commuting costs and other work-related consumer expenditures could be viewed as inputs to production rather than as outputs included in consumption (though it is not clear how these costs should be assigned for use in, say, productivity measurement). Conversely, some items now classified as intermediate inputs might better be classified as output for final demand. Researchers at the BEA have recognized this issue and changed the way they classify some market production. For instance, the BEA now classifies computer software purchases by businesses as investment rather than as an intermediate expense.

As with their market counterparts, nonmarket inputs and outputs must be properly classified for use in a double-entry accounting system and for the accounts to be useful for productivity analysis. Classification of market activities, much less nonmarket ones, is not always easy, but resolution of these classification issues will be a necessary step in the development of an expanded set of accounts.

Externalities

Where the state of theoretical modeling efforts and available source data permit, it would be extremely useful if satellite accounts included estimates of externalities associated with the covered activities. In this respect, satellite accounts would differ markedly from the NIPAs. An externality is an effect from the action of individuals or businesses that either damages or creates a benefit to others with no corresponding compensation paid to or received by those who engage in the activity. The treatment of externalities is a central issue for environmental accounting. The most important applications relate to air and water pollution. Whether failing to account explicitly for the potentially large negative externalities associated with pollution distorts measures of aggregate output depends on whether only market output or at least some nonmarket output is adversely affected. In the first case, aggregate output is measured correctly, though its allocation across sectors may be incorrect; in the second case, if pollution is considered a negatively valued product, even aggregate output is measured incorrectly (Nordhaus 2004). Positive externalities are important for evaluating investments in education, which may benefit not only the individual receiving the education but also the society at large, through having a better informed citizenry, lower crime rates, and so on.

It is easy to see how information about the magnitude of externalities might be helpful to policy makers. Knowing more about the externalities associated with air or water pollution would be valuable, for example, in determining the

[4]These views, as well as issues of classification more generally, are discussed in Conference on Research in Income and Wealth (1958).

level of taxes or permit fees for pollution emissions or waste disposal. Similarly, decisions about public investments in education would benefit from information about the externalities associated with a more educated population. It is equally clear, however, that accurate measurement of these externalities is apt to be a challenge.

Measuring Quantities

Dollar values are relatively easy to obtain for the market inputs to nonmarket production. If one wants quantity indexes for these market inputs, they can be constructed by using appropriate price indexes as deflators for the nominal expenditure data. In contrast, for both nonmarket inputs and nonmarket outputs, quantity measurement often will be a necessary first step in the development of monetary valuations.

Complications can arise even in the case of market inputs. Purchases of capital equipment by households, for example, are categorized as final consumption in the NIPAs. But measuring the inputs to household production requires an estimate of the stock of consumer durables. To create such a stock estimate, one must combine information on spending over time for dishwashers, refrigerators, vacuum cleaners, washing machines, and other capital equipment used in home production with information on these items' useful lives. Although there are practical difficulties that complicate estimation of the stock of capital equipment used in home production, the basic approach is well developed.[5]

An especially important nonmarket input on which, until very recently, quantity data have been lacking is the time devoted to nonmarket production. Fortunately, the American Time Use Survey (ATUS), launched at the start of 2003 by the Bureau of Labor Statistics, should go a long way toward filling this gap. The ATUS, described more fully in Chapter 2, can be expected to provide good data on the inputs of adult time to various sorts of nonmarket production in households of various types.

These data would be even more useful if the Census Bureau were to produce frequently updated information on the distribution of demographic characteristics in the population, designed to complement the new information on time and support accounting efforts generally. A complete demographic database might include information on the age, gender, school enrollment, years of education and degrees completed, occupation, household structure, immigrant status, employment status, and possibly other dimensions of the population. Knowing about the distribution of demographic characteristics and changes in that distribution over time would be of value, as an example, for determining whether observed changes

[5]This is a case for which the BEA already maintains a suitable data series, albeit not as a part of the core accounts. See Katz (1983) for a discussion of measuring the stock of consumer durables.

in the pattern of time use reflect changes in population mix or some other cause. The demographic data to support such an effort are, for the most part, already available, largely from the Census Bureau but in some cases from the Bureau of Labor Statistics, the National Center for Health Statistics, and other agencies. A determined researcher could compile these data from existing sources. But it would be very helpful if the information were assembled in a single place, adjusted to be consistent over time. The demographic database would not itself be a satellite to the existing economic accounts, but it would assist in the development and use of those accounts.

The ease with which the quantity of nonmarket outputs can be measured varies widely. Relatively good data are available, for example, on the educational attainment of the working-age population. These data provide a starting point for quantifying the output of the educational sector. Changes in mortality and morbidity are similarly well documented and could provide a basis for quantifying changes in the health status of the population, particularly if combined with demographic data tracking changes in population mix. In other cases, considerable creativity may be required to measure the quantities of nonmarket outputs, and doing an adequate job ultimately may require the collection of new data. Tracking air quality would require better measures of the pollutants to which the public is exposed and of the costs they impose. Tracking the output of the household sector would require data on such things as meals prepared or loads of laundry washed and dried. But, at least in principle, it is possible to see how this task might be approached.

To elaborate on the laundry example, the accounts would, on the input side, tally the number of hours devoted to laundry and the wage of a domestic employee or the opportunity cost or predicted market wage of the person doing the laundry (these methods are discussed below). The remaining inputs would be the capital services of the household's washing machine and dryer, together with the necessary materials, electricity, water, and detergent. These inputs would be reported in quantity and price terms. On the output side, the accounts would report the amount of laundry done and its price, estimated on the basis of what it would have cost to have the laundry done commercially. More thought needs to be given to what productivity measures mean when they are based on market substitute valuations. In the absence of direct measures of the output of nonmarket activities, one might impute them from observed market activities but, in such cases, productivity measures for nonmarket activities may simply recover the imputation scheme.

Assigning Prices

Anyone contemplating the development of nonmarket accounts must decide how best to value inputs and outputs in the various accounts, given the absence of prices. Valuation typically involves finding market analogues for the nonmarket inputs or outputs in question. Given the distance from the market of some utility-generating activities, however, this approach is not always feasible.

How to measure the value of unpaid time devoted to nonmarket production is the central input valuation issue. One approach has been to value this time at market substitute prices—the wage that would be paid to a person hired to perform the task in question. Another approach is to value time at the opportunity cost of the person performing the nonmarket activity. The two approaches may give quite different answers if higher-wage individuals devote time to tasks for which the market wage is relatively low.

At first blush, it may seem puzzling why anyone would choose to perform activities that compensate—in the form either of wages paid or value of nonmarket output produced—at a rate below the wage that could be earned in market employment. Further reflection makes clear that such decisions may be entirely rational. Economic theory conceives of people making marginal choices about their allocation of time to different activities. At the point of maximum satisfaction, the marginal earnings from an hour spent working for pay should be equated to the combined value of the output plus the value of the extra enjoyment intrinsic to an hour spent in nonmarket activity in comparison with market work.

A key point in this theory is the following: Even at the same moment, an individual may require different amounts of compensation, in the form either of dollars or the value of output produced, for engaging in different activities because different activities may yield different amounts of intrinsic satisfaction. A lawyer who commands $200 per hour from corporate clients may do work at $50 per hour for a charity. In this case, providing the work to the charity has an offsetting personal benefit (enjoyment) absent from working for a corporation. By the same principle, highly paid individuals may choose to prepare meals at home that could have been purchased in the market at a cost far below the wages the individual could have earned by working for pay instead of cooking. The recreation component of cooking means that the marginal value of the cooking performed is lower than the wage, if there is no similar recreational value in the person's job. In both cases—the lawyer performing work for a charity or the highly compensated person cooking meals at home—use of an unadjusted opportunity cost wage to value the time spent in activities the individual finds enjoyable would overstate the cost of inputs to nonmarket activities and understate their productivity.

We turn to economic theory for guidance in attaching an appropriate replacement cost value to time spent in nonmarket activities that someone else could have been hired to perform. The simplest case is that in which market and nonmarket production employ the same underlying technology. Consider a production function that relates productive inputs—labor *(L)* and capital *(K)*—to an output *(Q)*:

$$Q = f(b_i L, K). \tag{1.1}$$

Quantitatively, people's time (L) is the most important unmeasured input in nonmarket production. In the nonmarket context, an unpaid labor input must often be compared to a market replacement. People performing nonmarket tasks may be less skilled and work less hard on the task, on average, than people doing similar work in the market for pay. In the production function for nonmarket output, b_i is a measure of the relative efficiency of individual i in the nonmarket task as compared to the efficiency of market labor. If our speculation is correct, b_i will be a number between zero and one, though the precise value likely will vary across individuals. An appropriate procedure for cases in which a family member performs work at home that could have been performed by someone hired in the market is to count the family member's hours as measured and to value those hours at a rate equal to the efficiency factor, b_i, multiplied by the market wage for someone performing the type of work in question. Thus, if a homeowner chooses to reroof the house and, using the same materials and tools, takes twice as long to do so as it would have taken a professional roofer making \$30 per hour, we would record the time the homeowner spent on the task and value that time at \$15 per hour. Further, we would use the same \$15-per-hour valuation whether the homeowner earns \$100 or \$10 in his or her own market job. In the case of the \$100-per-hour person, we are implicitly assigning the roofing task an amenity value of \$85 per hour, while in the case of the \$10-per-hour person, we implicitly would be assigning it a disamenity value of \$5 per hour.

With respect to a task that cannot be given to another person—such as studying or exercising—there is little option but to use an opportunity cost valuation of the time expended. For people who work in the market, the market wage rate commonly is used as a measure of opportunity cost; for those who do not, a market wage must be imputed. Having said this, it should be recognized that people perform activities at home at different times of the day, week, and year than those they perform in the market. The marginal wage rate that a person might obtain in the market at the times many household activities are performed is unlikely to equal the average wage rate the person receives for his market activities. For example, the relevant hours for home production may be the 2-3 hours at home after putting in a full workday at a market job. It is not clear that the market value of those hours is the same as the individual's marginal or average hourly rate. Furthermore, many people in the labor market are unable to sell additional hours at their wage rate. Wage workers have little control over their hours; salaried workers, by virtue of the compensation structure of their jobs, are typically not paid at all for extra hours of work (Nordhaus, 2004, p. 18). For these reasons, the market wage rate may be a poor measure of the opportunity cost of an hour of the person's time. Recognizing this problem is easy; doing anything about it is difficult. Certainly, more research into the relationship between opportunity cost and market wages would be helpful here.

More fundamentally, even beyond the difficulty of measuring the opportunity cost associated with time devoted to nonmarket production, one would want in principle to adjust that opportunity cost to reflect differences between the amenity of work and the nonmarket activity in question to estimate the true cost of time devoted to the latter. Making this adjustment would require being able to quantify the value of enjoyment that individuals derive from different activities.

Provided that individuals are able to equalize the marginal value of time across all of the activities in which they engage, the adjusted opportunity cost measure we advocate for valuing tasks that someone else cannot be hired to perform is conceptually equivalent to the adjusted replacement cost measure that seems most appropriate for valuing time devoted to nonmarket activities satisfying Reid's third-party criterion. To see why this is so, return to the roofing example discussed above. Let W_S equal the hourly wage of a roofing specialist (replacement wage) and W_{Oi} the foregone hourly market earnings (opportunity wage) of a homeowner taking on a roofing task. The productivity-adjusted replacement cost of time the homeowner devotes to roofing would be

$$\hat{W}_{Ri} = b_i W_S \tag{1.2}$$

where, as before, b_i is a parameter between zero and one that reflects the relative productivity of the homeowner as compared to the professional in the activity under consideration. The amenity-adjusted opportunity cost of time the homeowner devotes to the activity—in this case, roofing—rather than to paid employment would be

$$\hat{W}_{Oi} = W_{Oi} - e_i \tag{1.3}$$

where e_i represents the value of the extra enjoyment that individual i receives from spending an hour in the nonmarket activity rather than in paid employment. But for any given activity, an optimizing individual will seek to equate the value of the marginal product associated with time devoted to an activity ($b_i W_S$) to the net opportunity cost of that time ($W_{Oi} - e_i$).

This discussion implies that there are two alternative but potentially equivalent routes to arriving at a valuation for time devoted to nonmarket activity. If we know the relative efficiency of the family member, the effective wage can be calculated as the professional's wage multiplied by that relative efficiency. We could estimate the amenity value the person attaches to the nonmarket work by subtracting the effective wage from the individual's market value of time. Or, if we know the amenity value of the activity to the individual, we can subtract it from the family member's market value of time to estimate the effective wage. As a byproduct, at least in the case of activities that could have been performed by a third party so that an observable wage exists for market providers, we could estimate the person's efficiency relative to a professional in the same line of work as the ratio of the effective wage to the professional's wage. Where it is an option, it seems likely that adjusted replacement cost will be easier to estimate than adjusted opportunity cost, but conceptually the two approaches should lead to much the same result.

Additional factors, such as taxes, complicate the relationship between opportunity and replacement costs. All else the same, income taxes reduce the opportunity cost of a person's own time devoted to nonmarket production, but not the replacement cost of hiring someone else to perform the task. This suggests the use of pre-tax wage rates in replacement cost calculations but after-tax wage rates as the basis for opportunity cost calculations. Still, the basic concepts to be applied are reasonably straightforward.

Transaction costs may be an important consideration in certain cases. A homeowner who hires a roofer does not entirely escape the need to spend time on the project—time must still be devoted to finding and negotiating with the roofer and to overseeing the roofer's work. Incurring these costs may be worthwhile for a large job but not for a small job. Market service providers also may incur transactions costs that make it unattractive for them to take on small jobs. It would be very difficult, for example, to find anyone willing to come to your house for 10 minutes each morning just to prepare a pot of coffee!

We should also acknowledge that things are more complicated when different technology is employed for market and nonmarket production. Continuing with the example above, a professional roofer might find it worthwhile to purchase specialized tools and equipment that a homeowner doing the same job would not possess. Even if skill levels and intensity of work effort were similar, this might be another reason to expect that a professional roofer could complete a job more quickly than the average homeowner. The homeowner's decision between hiring a roofer and doing the work personally will depend on the full relative costs, not just on the time requirements, associated with the two options.

In many if not most cases, however, market and nonmarket producers perform their tasks in similar ways—doing yard work, ironing shirts, and making a salad for dinner come to mind as examples. In such cases, the only significant difference between market and nonmarket production is the source of the labor supplied and, even in cases where other factors come into play, it is likely to be the most important difference.

Recommendation 1.4: A replacement cost measure, adjusted for differences in skill and effort between nonmarket and market providers, should be adopted for valuing time inputs devoted to nonmarket production in cases in which someone else could have been hired to perform the work in question. Where this is not the case, an opportunity-cost-based measure, ideally adjusted to account for the intrinsic enjoyment associated with the activity in question, should be used for valuing time devoted to nonmarket production activities.

We believe that the valuation of nonmarket *outputs* should, where possible, follow the principle of treating nonmarket goods and services as if they were produced and consumed in markets. Under this approach, the prices of nonmarket goods and services are imputed from a market counterpart. Many youth sports organizations, for example, are operated largely by volunteers. Although a fee may be charged for participation in the activity, that fee cannot be viewed as a market price. But there are also private firms that offer opportunities for children to participate in similar recreational activities that do charge a market-determined price. Given information on the relevant output quantities—for example, the numbers of children participating in the various recreational programs of a nonprofit youth sports organization—the price charged for participating in similar activities offered by private firms could be used in valuing the nonprofit organization's output.

In some cases, there may be differences in quality between home-produced outputs and market outputs, just as there may be between home and market production inputs. In principle, the valuation of nonmarket outputs should take into account any differences in the quality of those outputs as compared to similar market outputs, much as we proposed for the valuation of nonmarket as compared to market labor inputs.

Even in the case of near-market goods, market and nonmarket outputs may be imperfect substitutes, complicating comparisons of their value. More difficult yet are the cases in which a nonmarket good is an asset that has no direct market counterpart and is never sold. A possible approach in these cases may be to use market prices to value the stream of output produced by the asset over time and then to treat the present value of the returns as a measure of the asset's value. This approach has a clear grounding in the standard theory that underlies the valuation of marketable capital assets and is the approach taken, for example, by Jorgenson and Fraumeni (1989, 1992) in their work on the valuation of investments in human capital. They begin by calculating the increments to earnings associated with successive increments to education. The present value of the earnings increments, cumulated over a person's productive lifetime (and assuming that education enhances the value of market and nonmarket time equally), then is used as a measure of the value of the incremental investment in human capital.

Investments in health also yield a flow of nonmarket services over time. Improved health increases not only expected years (and quality) of labor market activity, and thus labor market earnings, but also the expected number of years available in which to enjoy all that makes life rewarding. Developing a market-based measure of the marginal value of additional years of life that may flow from health care investments is controversial, though labor market data have proven useful for this purpose. Specifically, the fact that different occupations are associated both with different risks of fatal injury and different relative wage rates has been exploited to derive estimates of the value of an additional year of life. Such measures, while far from perfect, have the advantage of being based on real-world decisions that yield observable market outcomes and, for that reason, they have appeal.

Different approaches may be necessary for the case of nonmarket outputs that are public in nature, such as crime rates and air quality. Again, however, it may be possible to develop measures of the value of these outputs on the basis of market transactions. The levels of many, if not all, of these nonmarket outputs are likely to differ across localities. People presumably will be willing to pay more to live in communities with low crime rates and good air quality than in communities that lack these attributes. The value of such positive attributes are reflected in house prices. At least in principle, one could derive an estimate of the value of lower crime rates, better schools, or higher air quality from a hedonic model that relates house prices to these (and other) community characteristics.[6]

There are a number of areas for which market valuation, or even imputations based on nonmarket analogues, are simply unavailable and impossible to obtain. Examples of these might include some aspects of social capital, such as degree of social cohesion or cooperation; the effect of terrorism on the population's sense of well-being; or the "existence" and "legacy" values of national monuments, such as the Grand Canyon. In these cases, any attempted valuation would have to rely on more indirect evidence. As indicated above, we would argue strongly that attention should be directed first to those categories of nonmarket output for which the most defensible, market-based approaches to valuation are possible.

Counting and Valuation Issues

The national accounts have a consistent structure for reporting prices and corresponding quantities. The two have an intimate connection because prices form the basis for aggregating the quantities of different products. The national accounts have adopted the approach long advocated by index-number theorists—chain-weighted quantity indexes of groups of products are computed by weighting the percent change of the quantity of each product by its share in the dollar value of all the products. As a result, the accounts support productivity calculations directly. Productivity growth for any group of products—including all the products in GDP—is the percent growth of the aggregate quantity less the corresponding weighted growth of the inputs.

In the market economy, monetary aggregates generally are the most accessible measures of the level of activity—dollar values of sales, dollars paid as wages and salaries, and so on—and measuring quantities often is more difficult. By definition, however, nonmarket activity does not involve monetary transactions. Consequently, data on monetary aggregates that form the building blocks

[6]See Heckman and Vytlacil (1999) and Heckman et al. (2003) for an indication of advances in the theory of hedonic price measurement in this context.

for traditional national income accounting are simply not available. Instead, available data may consist of physical or other quantity indicators of the level of activity, such as hours of time devoted to home production, student-years of education provided, or ambient concentrations of various air pollutants.

On one side are those who argue that no nonarbitrary way exists for assigning monetary values to a heterogeneous set of nonmarket inputs or outputs, and that any such assignment unavoidably will reflect value judgments that are inappropriate for a statistical agency (see, e.g., van de Ven et al., 1999, p. 8). The counterposition holds that, without an attempt to assign monetary values to the quantity indicators that are the basic unit of measurement for nonmarket outputs, it will be difficult for policy makers to digest and use the information (see, e.g., National Research Council, 1999, p. 123). This may mean that nonmarket outputs end up being ignored, which implicitly assigns them a value of zero. Alternatively, policy makers may assign a value to the nonmarket output using subjective methods that are less defensible than the methods that would be employed by a statistical agency. In either case, there is a good argument for measurement specialists to provide estimates based on the best possible methods, even if these are highly imperfect, rather than leaving a statistical void. Another argument for attempting to assign monetary values to quantity indicators is that the effort filters out indicators that may be of minor economic importance. One problem with purely physical accounting systems is that, useful as they may be for some research topics, they tend to be encyclopedic and difficult to comprehend. Economics can minimize biased value judgments by providing scientific guidelines and accounting conventions for approximating prices in many cases. And with a monetary metric, the aggregation of detailed measures of output to larger, useful indexes is possible. For these reasons:

> **Recommendation 1.5:** Wherever feasible, nonmarket inputs and outputs should be valued in monetary, not just physical unit, terms.

The usefulness of a monetary valuation approach depends on the extent and accuracy with which monetary values ultimately can be assigned to the inputs and outputs in question. In order that such assignments be as objective as possible, we favor basing these valuations wherever possible on information derived from the terms of observable market transactions or their analogs. Even when it is difficult to base valuations on market transactions, it is important that calculations be, in principle, reproducible by independent observers. In certain instances, assigning prices to outputs (or inputs) may be so controversial that publishing physical quantity accounts may be the best available option. Given that both price and quantity data are needed to calculate values for the conventional monetized accounts, however, it is reasonable to produce the best price and monetary estimates available (along with the quantity data), as long as sets of assumptions are clearly stated. Limiting an account to physical quantity reporting should be the exception, not the rule. We also again emphasize the desirability of giving priority in the near term to work in those areas of nonmarket accounting for which valuation can draw from market comparisons.

Marginal and Total Valuation

Economic valuation methods fall into two broad categories: the first, which tracks the framework of the national accounts, relies on prices, which reflect marginal benefits; the second considers the full amount consumers would be willing to pay for a good or service, which includes a consumer surplus to the extent that amount is greater than the price. Thus, the two approaches differ in the way benefits are measured.

In the case of a product or service sold in a competitive market, the price is set at a value that equates the cost of producing and the value of consuming the marginal unit of output. In a standard supply-and-demand diagram, shown in Figure 1-1, a value of P^* is assigned to each of the Q^* units of output, for a total market valuation represented by the shaded rectangle. The total value to the consuming public of these Q^* units of output, however, is the larger, total shaded area. The difference between these two areas, the shaded triangle, is the consumer surplus associated with the consumption of Q^* units of the product or service in question at a price P^*.

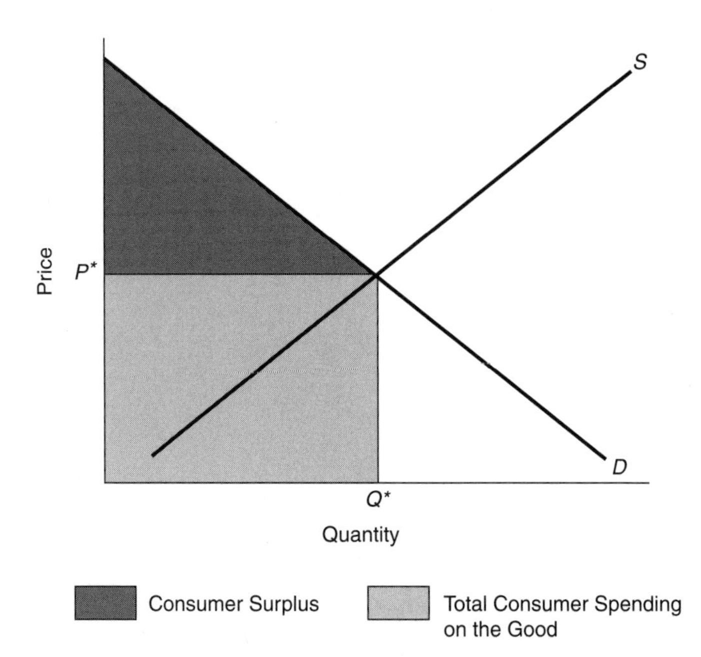

FIGURE 1-1 Consumer surplus.

In many cases, knowing consumers' willingness to pay for first and subsequent units of a good or service does not matter for any decision. Although the public enjoys a large consumer surplus from the production of ice cream—that is, enjoyment exceeding in value the total price paid for the ice cream—there is no policy or accounting issue relating to that surplus. Productivity and other types of measurement use the marginal values revealed by the market price. The same principle applies to many of the nonmarket goods and services that would be included in satellite accounts.

In the context of national accounting, the potential usefulness of total valuation seems to us to be restricted to the case of products new to the market where it may be important for accurate comparisons of the value of output in the current period as compared to that in prior periods. It has been argued that the value consumers place on new products should be reflected in properly constructed price indexes as a decline in the price level (see, for example, Hausman, 1996). While there is not yet a consensus on this issue in the price index literature, we would note that deflating nominal expenditures with a price index that accounted for the value realized by the purchasers of new goods would yield an estimate of real output that included consumer surplus associated with the introduction of these goods. It is meaningless, in a national income accounting context, to estimate total value for existing products. Sometimes total value data will be needed for a cost-benefit analysis, and this is fine; cost-benefit analysis and national accounts rest on different conceptual ideals and objectives.

Recommendation 1.6: As a general rule, nonmarket accounts should measure the marginal valuations associated with covered goods and services.

2

Accounting and Data Foundations

A large portion of this report is devoted to working through the many conceptual challenges that clutter the path toward construction of nonmarket satellite accounts. Another obstacle to this achievement, and one that will be expensive to clear, is the lack of data to support quantification and valuation of nonmarket activities in an acceptable fashion. As noted above, the new American Time Use Survey (ATUS) will provide rich information on the most important input to nonmarket production—the time people devote to nonmarket activities. Other inputs to nonmarket production commonly are purchased in markets, so that the challenges associated with their measurement, while not trivial, should be similar in nature to those routinely encountered in the construction of the National Income and Product Accounts (NIPAs). Considerable work will be required to develop the data needed for independent measurement of nonmarket outputs.

In this chapter, we begin, for reference purposes, with an overview of the NIPAs—the prominent U.S. model for national economic accounting. Many of the principles used in the construction of the NIPAs also carry over in a natural way to the construction of nonmarket economic accounts. We next discuss the role of time-use data in nonmarket accounting and describe the new ATUS. Another cross-cutting data development need is that for a coherent, readily accessible demographic database. The data needed to support measurement of nonmarket outputs are diverse; development of these output measures will be the house-to-house combat of nonmarket accounting.

OVERVIEW OF THE NATIONAL INCOME AND PRODUCT ACCOUNTS

The purposes of this section are to present an overview of the NIPAs, the standard to which we refer often throughout this report, and to provide a basis of comparison for concepts and data needs for the nonmarket accounts described in subsequent chapters. National income and product accounting is the centerpiece of national economic accounting in the United States. The NIPAs show the real and nominal value of output, the composition of output, and the distribution across types of income generated in its production.

There are three other major branches of national economic accounting—capital finance accounting, balance sheet accounting, and input-output accounting. The capital finance accounts are better known as the flow-of-funds accounts. They show the role of financial institutions and instruments in transforming saving into investment and the associated changes in assets and liabilities. These changes occur as monetary flows over time, resulting in an increase or depreciation in the stock (the accumulated amount) of the asset. Balance-sheet accounts display these assets and liabilities at particular points of time. Input-output accounts trace the flow of goods and services among industries in the production process, and show the value added by each industry and the detailed commodity composition of output. Input-output matrices, particularly the benchmark input-output matrices, are a foundation for the NIPAs. Other accounts, specifically the international and the regional accounts, are also important sources of information for the national accounts (Bureau of Economic Analysis, 1985). These other accounts are mentioned only briefly in this chapter.

In the NIPAs' double-entry accounting system, domestic output can be measured by either gross domestic product (GDP) or gross domestic income (GDI). GDP is measured as the market value of goods and services produced by labor and property located in the United States. GDI is measured as the costs incurred and the incomes earned in the production of GDP. Business purchases from other businesses are netted out so that domestic output is an unduplicated total. In theory, nominal GDP should equal nominal GDI; because the two sides of the accounts are measured using independent and imperfect data, however, the aggregates typically do not match. The statistical discrepancy, which is recorded as an income component, is equal to nominal GDP less nominal GDI. Within a production function framework, GDP represents the economy's output and GDI represents the capital and labor inputs used in its production, plus taxes on production (such as sales and excise taxes) and the surplus of government enterprises.

In the United States, GDP is measured using an expenditure approach. GDP, as shown in Table 2-1, is equal to the sum of personal consumption expenditures, gross private domestic investment, net exports of goods and services (exports minus imports), and government consumption expenditures and gross invest-

TABLE 2-1 Gross Domestic Product and Gross Domestic Income, 2001 (billions of dollars)

Compensation of employees, paid	5,881.0	Personal consumption expenditures	6,987.0
Taxes on production and imports less subsidies	659.8	Gross private domestic investment	4,109.9
Net operating surplus	2,329.3	Net exports of goods and services	–348.9
Consumption of fixed capital	1,329.3	Government consumption expenditures and gross investment	1,858.0
Gross Domestic Income	10,199.4		
Statistical discrepancy	–117.3		
Gross Domestic Product	10,082.1	Gross Domestic Product	10,082.1

SOURCE: Mayerhauser et al. (2003, p. 10).

ment. The change in private inventories is included in the gross private domestic investment component of GDP. Imports are subtracted in the calculation of GDP, as expenditures for both consumption and investment include imported goods and services that are not a part of domestic output.

GDI is equal to the sum of compensation of employees, taxes on production and imports (minus subsidies), net operating surplus, and consumption of fixed capital; a close synonym for consumption of fixed capital is economic depreciation. Operating surplus is one of several new concepts introduced into the NIPA framework as part of the December 2003 comprehensive revision. Net operating surplus is a profits-like measure that shows business income after subtracting the costs of compensation of employees, taxes on production and imports (minus subsidies), and consumption of fixed capital from gross product (or value added), but before subtracting such financing costs as net interest and business transfer payments (Mayerhauser et al., 2003). GDI does not include capital gains on assets as these do not arise from current production. Consumption of fixed capital is an entry that represents a cost of production, as it is a charge for capital used up during the accounting period.

The double-entry nature of the NIPAs is not obvious from Table 2-1, as there are entries with different titles under GDP and GDI, but if an entry appears on one side of the accounts, a corresponding amount must appear someplace on the other side. For example, if an individual buys an automobile with his wage income, this amount appears as labor compensation under GDI and as a personal consumption expenditure under GDP. If the individual saves part of his wage income, that saving appears as investment under GDP. By definition, gross investment is equal to gross saving plus the statistical discrepancy, though the relationship between GDI and GDP is less obvious in the savings case than in the automobile purchase case.

Although nominal GDP should in theory be equal to nominal GDI, there is no expectation that, in practice, real GDP will equal real GDI. In fact, multifactor productivity is typically estimated by the difference in the rate of growth of output compared to the rate of growth of inputs. The NIPAs as currently formulated do not allow direct estimation of multifactor productivity, but the Bureau of Labor Statistics estimates multifactor productivity by major sector largely with data from the NIPAs (Fraumeni et al., 2004) and the international System of National Accounts (Commission of the European Communities et al., 1994) may be modified to call for the estimation of capital inputs within a national income accounting framework.

Beyond pure monetary-based calculations, the NIPAs include a number of imputations, amounting to approximately 15 percent of GDP (Moulton, 2002). Although it is difficult to define rigorously what an imputation is, Moulton (2002, p. 3) offers the following:

> An imputation in national accounts refers to a flow that must be estimated by the national accountant because there is no directly related monetary transaction that is recorded in the books of a party to the transaction. Imputations generally arise for one of two reasons: (a) own-account production that takes place within the production boundary of the system, such as the services produced by owner-occupied dwellings, or (b) transactions that are not directly associated with an exchange of money between the transacting parties because the transactions involve barter, transactions in kind, or bundling the provision of a service with a financial transaction, such as depositing funds in a bank.

As Table 2-2 shows, four categories account for the majority of the value of imputations in the NIPAs: services provided by owner-occupied housing, which accounts for about 6 percent of GDP; employment-related imputations, which account for about 3 percent of GDP; unpriced services provided by financial

TABLE 2-2 Imputations in the NIPAs, 2001 (billions of dollars)

Gross domestic product	10,082.2
Imputations	1,549.5
Owner-occupied housing	648.5
Rental value of nonresidential fixed assets owned and used by nonprofit institutions serving individuals	61.2
Services furnished without payment by financial intermediaries except life insurance carriers	288.9
Employment-related imputations	354.8
Farm products consumed on farms	0.2
Margins on owner-built housing	8.2
Consumption of general government fixed capital	187.7
Excluding imputations	8,532.6

SOURCE: Moulton (2002, p. 11).

intermediaries, which account for about 3 percent of GDP; and services from general government fixed capital, which amounts to something less than 2 percent of GDP.

This outline of the NIPAs focuses on measures for the economy as a whole. Sectors covered in the accounts include the business, household, and government sectors. The Bureau of Economic Analysis (BEA) also has produced tables showing information for nonprofit institutions serving individuals, which represent the vast majority of nonprofits; in the NIPAs these nonprofits are lumped together with households. Acknowledging the imputations in and sector coverage of the NIPAs highlights the fact that the accounts do not exclude all nonmarket activities or the services from all nonmarket assets. Many government and nonprofit activities register in the NIPAs but, in general, only goods and services that involve payments (whether market or below market) are included. For example, payments for services rendered under the Medicare program are included under GDP and GDI, but no value is imputed for volunteer services in nonprofit hospitals, because they are provided without remuneration. Services from owner-occupied housing and from general government capital are imputed and included in the NIPAs. Valuation of the services from owner-occupied housing is linked to the terms of observed rental transactions. BEA acknowledges that the services from general government capital are underestimated in the accounts; uncertainty about the appropriate net return to such capital led to the adoption of an assumption that the net rate of return was zero (National Research Council, 1998; Bureau of Economic Analysis, 1995).

In addition to the summary domestic income and product accounts shown in Table 2-1, the NIPA summary accounts include six others: the private enterprise account, the personal income and outlay account, the government receipts and expenditures account, the foreign transactions current account, the domestic capital account, and the foreign transactions capital account. Together these summary accounts, from which approximately 150 NIPA tables and the fixed-asset estimates (which underlie NIPA measures of consumption of fixed capital and consumer durable investment) are derived, provide a wealth of information for users (see BEA website).

The satellite accounts envisioned in this report would build primarily on the domestic income and product accounts, which include information on the production costs that constitute GDI and on the values of final output that constitute GDP. In the satellite accounts we contemplate, the nonmarket analogs are values for the inputs used in nonmarket production and values for the output that is produced.

MEASURING TIME USE

Time is the most quantitatively significant input to both market and nonmarket production. One cannot begin to understand economically oriented nonmarket activity without knowing how a population spends its time. Data appropriate for measuring time devoted to nonmarket activities require the recording of information on people's activities away from their jobs. Just as market transactions are recorded directly by measuring expenditures rather than relying on people's recollections of what they produced or purchased, it is also important to measure what people actually do with their time, not what they recollect having done with it long afterwards. The vehicle for collecting such information is a time-budget survey—a study in which a large sample of individuals keep a diary of their activities over one or several days. In a time-budget survey, respondents describe the various activities in which they have engaged and these are then coded into a set of categories. A well-designed time budget survey forces the aggregate of time devoted to all activities to equal 1,440 minutes per day for each person.

Previous Collections of Time-Use Data

Most classifications of household time use have followed the precedent set by Sandor Szalai (1973) and his collaborators, who organized time-use surveys in the 1960s to record people's activities in a number of countries. Time budgets have been collected in the United States since at least the 1930s (Sorokin and Berger, 1939). There have been a few American time-budget surveys in the post-World War II period, including surveys in 1965-1966 and 1975-1976 (with a small extension in 1981), conducted at the University of Michigan (Juster and Stafford, 1985). Time-budget surveys were conducted at the University of Maryland in 1985, 1992-1994, and 1995 (Robinson and Godbey, 1999); a more recent study focusing particularly on child care activities also was done at the University of Maryland (Bianchi, 2001).

While all of these studies were funded by federal agencies, none was designed or conducted by any part of the federal statistical apparatus. There have been differences in sample design across the surveys, in the ages of respondents to which the sampling frame applied, in the number of diary days sampled per individual, in the number of activities categorized, and in the categorization of activities. Perhaps most important for our purposes, the samples have typically been quite small—never more than 8,000 diary-days (1992-1994)—and they have been conducted on an irregular basis.

Internationally, interest in time-use studies as a tool to estimate the quantity and, in turn, the value of nonmarket work has grown considerably over the last 20 years, and a number of countries are now committed to this task. In 1995, the U.N. Fourth World Conference on Women called for national and international statistical organizations to measure unpaid work and estimate its value in satellite accounts to the GDP. Other countries' statistical agencies have conducted much larger-scale time-budget surveys in recent years, albeit only on a quinquennial or decennial bases, with several using the resulting data in the construction of nonmarket accounts. Efforts to date that are most directly relevant to the United

States are those underway in three other large English-speaking countries—Australia, Canada, and the United Kingdom.

Canada first administered a national time-use survey in 1981 and subsequently established it as a regular component of its General Social Survey. Additional questions on unpaid housework and child care and elder care were asked on the 1996 census. Statistics Canada has used time-budget data to construct estimates of the value of households' unpaid work. The most recent estimates, constructed using a replacement cost approach, put the value of households' unpaid work in 1992 at about 34 percent of GDP (Statistics Canada, 1995, p. 42).

Australia administered a pilot national time-use survey in 1987, with expanded versions in 1992 and 1997. The Australian Bureau of Statistics (ABS) published estimates of the value of unpaid work based on the 1997 survey. Relying on several different measures of replacement cost, ABS calculated that unpaid work amounted to about 48 percent of Australian GDP in that year (Australian Bureau of Statistics, 2000, p. 5). The divergence in the Canadian and Australian estimates of the amount of unpaid work likely reflects differences in the methodologies used in their time-use surveys, including differences in the categorization of activities, but no detailed comparisons or explanations have been offered to date.

The United Kingdom administered a time-use survey in 2000, but in developing measures of nonmarket output the U.K. Office for National Statistics has applied an output-based method of valuation, rather than simply assigning a replacement value to time devoted to unpaid work (Holloway et al., 2002). The experimental accounts that have resulted are focused on several different outputs of the household sector: provision of housing, transport, nutrition, clothing and laundry, child care, adult care, and volunteer activity. Perhaps because of the emphasis on methodological development, the office did not provide an estimate of the size of the country's household sector relative to standard measures of GDP (Holloway et al., 2002).

Pioneering academics have engaged in lonely but important data collection efforts to collect time-budget data for the United States. Mainly because of the absence of continuing federal funding for this activity, however, it seems fair to conclude that the United States has until very recently been in the *derrière garde* worldwide in the collection of such data.

The American Time Use Survey: Overview

In January 2003, the Bureau of Labor Statistics initiated the monthly American Time Use Survey (ATUS). This study originated in part out of research interest in valuing women's time in the household. Concerns that women's contributions were being undervalued by the exclusion of household production activities prompted the initial Bureau of Labor Statistics (BLS) efforts to develop and test the collection of time-use data, leading to a pilot study in 1997 and full-scale field testing in 2002 (see Horrigan and Herz, 2005). Once estimates have been subjected to appropriate scrutiny and reasonably well verified, the data from this survey will be a crucial input into the creation of nonmarket accounts.

> **Recommendation 2.1:** The American Time Use Survey, which can be used to quantify time inputs into productive nonmarket activity, should underpin the construction of supplemental national accounts for the United States. To serve effectively in this role, the survey should be ongoing and conducted in a methodologically consistent manner over time.

The sampling frame for the ATUS is that of the monthly Current Population Survey (CPS)—the actual ATUS samples are taken randomly from households just completing their eighth month in the CPS sample. For example, a household that had been included in the CPS in January through April 2002 (waves 1-4) and January through April 2003 (waves 5-8) might have been included shortly thereafter in the ATUS (a new wave 9). Households are chosen based on a variety of stratifications (including race/ethnicity and the presence of children of various ages), all designed to reduce the sampling variance of the time-use statistics that cover smaller subsets of the U.S. population.

One randomly selected (by BLS) adult member in each household chosen for participation in the ATUS is asked to complete a time diary. The diary is to be completed for the previous day, with a telephone interviewer leading the respondent through his or her activities over the 24-hour period that began at 4 a.m. on that day. Ten percent of the diary days are assigned to each weekday (Monday through Friday), 25 percent are assigned to Saturday, and 25 percent are assigned to Sunday. Respondents list their activities, showing when each new activity began and describing it in their own words. "Secondary" activities, undertaken simultaneously with the listed activities, are recorded if the respondent volunteers that they occurred. The respondent also lists where each activity was undertaken (e.g., at home, at the workplace, elsewhere) and who else was present (e.g., nobody else, spouse/partner, child/children, friends, coworkers).

A crucial issue for our purposes is the classification of the respondents' verbal descriptions of activities into categories that are useful for accounting and analysis. While the coding system created by the Szalai group has underlain the sporadic U.S. time-budget surveys, the ATUS has gone far beyond this. Beginning with a three-tier six-digit coding system, the basic codes are aggregated into 17 top-level categories:

- personal care activities (mainly sleep);
- household activities;
- caring for and helping household members;
- caring for and helping non-household members;
- work and work-related activities;
- education;

- consumer purchases (e.g., food shopping);
- purchasing professional and personal care services (e.g., doctors' visits);
- purchasing household services;
- obtaining government services and civic obligations;
- eating and drinking;
- socializing, relaxing, and leisure;
- sports, exercise, and recreation;
- religious and spiritual activities;
- volunteer activities;
- telephone calls; and
- traveling.

This categorization appears to accord well with the construction of supplemental accounts along the lines discussed in this report.

In addition to completing the time diaries, ATUS respondents again answer most of the questions that they had been asked in the CPS, providing updates on their work behavior, demographic characteristics, earnings and (bracketed) family income. These additional data keep economic information current, which is important for allowing the full ATUS sample to be used in valuing time in household production and for more accurate construction of estimation weights.

BLS had expected to sample roughly 2,800 households per month in 2003 and to obtain a 70 percent response rate. But the response rate from the diaries taken by telephone was only 59 percent, and the response rate for the small number taken in person (from households without telephones) was only 34 percent. Because some of the 2003 sample was supported with funding that had been carried over from the survey development period, fewer households can be surveyed on a continuing basis. About 1,800 households per month were to be surveyed beginning in January 2004, with actual responses expected from individuals in about 1,200 households. Thus, the number of individual observations available for 2003 is about 21,000; roughly 14,000 individuals are expected to complete diaries in 2004 and each year thereafter. Significant nonresponse in a sample already trimmed by dropouts from participation in the CPS certainly has the potential to adversely affect the quality of the ATUS data, a matter to which we return below.

As a large-scale and on-going time-budget survey, the ATUS is unique worldwide. Several other countries' time-budget data sets are large enough to generate reliable measures of time allocation and include enough economic and demographic information to allow values to be attached to the hours they spend in productive activities. Other countries thus can construct statistically meaningful point-in-time supplemental national accounts, and some have done so. No other country, however, currently has the ability to produce satellite nonmarket accounts that can be continuously updated. The size of the underlying samples in the ATUS soon will be the largest in the world, but what makes the survey particularly valuable for the purposes of creating regularly published nonmarket accounts is that its information will be provided every year, not just at intervals.

The American Time Use Survey: Problems—and Nonproblems

During its deliberations, the panel heard from a number of people who have been involved in the development and fielding of time-use surveys. Several of them raised questions about certain features of the ATUS design, questions which raise legitimate concerns. Not all of these concerns, however, are directly relevant to the potential usefulness of the ATUS for constructing nonmarket economic accounts.

Despite the tremendous step forward that the ATUS represents, and its prospective role as a linchpin of nonmarket economic accounts for the United States, there are real concerns about its reliability for this purpose that need to be addressed. They can be summarized by the questions: Who? When? and What? The most important, the "who" concern, is engendered by the lower-than-expected response rates in the ATUS. Response rates on the precursor studies also have been low (Egerton et al., 2004), but the 59 percent response rate on the ATUS was even lower, and far below the 90-plus percent response rates in the CPS.

Because most of the ATUS nonrespondents provided CPS responses the month before, a good deal is known about their demographic and economic characteristics. It is a relatively simple matter to reweight the sample averages of time allocations to account for differential nonresponse rates across groups with different observable characteristics. The difficulty is that there is no reason to assume that nonresponse is random (relative to the CPS sampling frame) across unobservable characteristics that may be correlated with time allocations. Although there is no way to know for sure until the data can be carefully examined, it seems plausible that busier individuals, or those with more irregular schedules, might simply be less likely to participate in the survey, meaning that the survey estimates could be distorted. Additionally, these biases in the ATUS are compounded with any biases in the sampling frame related to survivorship from the CPS.

There is no simple way to adjust for nonrandom nonresponse related to unobservables. The BLS is fully aware of this difficulty and has explored various means of boosting survey response rates (Horrigan and Herz, 2005). Nonetheless, low response rates—and the possible resulting nonresponse bias—are the biggest concern with using ATUS data to construct satellite nonmarket accounts. Efforts to assess the extent of any possible bias in the survey responses—and, if necessary, to address that bias by raising response rates or making appropriate adjustments to the estimates—should be a priority.

Recommendation 2.2: The Bureau of Labor Statistics should commit resources adequate to improve response rates in the American Time Use Survey and to investigate the effects of lower-than-desirable response rates on survey estimates.

The "when" question stems basically from the nonresponse problem, and it has two aspects. First, are the ATUS responses representative of activities in which people engage on the days of the week for which they complete time budgets? As with the "who" question, it is a simple matter to reweight completed diaries so that each day of the week receives the same weight in the aggregated survey estimates. But even if BLS received diaries from an apparently (based on observable characteristics) random sample of individuals across days of the week, one cannot know whether the nonrespondents are distributed differently by day of the week in relation to their unobservable characteristics. One might, for example, believe that those whose time is relatively more valuable on weekends, given their observable characteristics, would be less likely to respond if asked to complete a time budget for a Saturday or Sunday. Difficulties with response rates generate difficulties with inferring the true weekly distribution of time allocations from the distribution of completed time-budgets. This underscores the importance of Recommendation 2.2.

A second "when" issue is seasonal and is qualitatively similar to the possibility of nonrandom response propensities by day of the week: Might nonresponse rates across weeks of the year also be nonrandom in the unobserved characteristics of the individuals who are sampled? Among observationally identical individuals who are sampled in, say, July, those who respond may tend to be those with relatively less active schedules. Again, there is a risk that response probability may be correlated with patterns of time allocation. Raising response rates also is likely to reduce this problem.

The "what" question of particular relevance to the panel's purpose is the extent to which the ATUS records activities that are performed simultaneously (e.g., secondary, multiple, or standby activities).[1] BLS has made a concerted effort—far more intensive than in the precursor studies—to ensure uniformity in the coding of respondents' descriptions of their primary activities and has created the most detailed set of basic codes ever used in a time-budget survey. One limitation of the data produced by the survey is that they track "primary" activities, but not secondary ones; in other words, the data are coded to show people engaged in just one activity at a time. The survey does include separate questions designed to learn about time devoted to child care activities, which empirically is by far the most important "secondary" activity reported by respondents in other time-use surveys. Still, more complete information about secondary activities could prove to be important for monitoring time devoted to productive nonmarket activities that may occur simultaneously with other tasks or pastimes. A related

[1]The issue of standby and secondary activities is thoroughly discussed in Pollak (1999).

question is whether activities that typically require only a few minutes at a time—for example, moving a load of laundry from the washer to the dryer—will be reported consistently enough to support good estimates of time devoted to them.

Cost and burden considerations led BLS to decide against trying to elicit information about secondary activities generally; it records them only if the respondent volunteers that he or she was engaged in something else simultaneously with the primary activity, and it does not now code this information. While BLS is examining potential ways to obtain more information on secondary activities, the present lack of comprehensive information about them is a significant limitation for using the data to construct nonmarket accounts.

Recommendation 2.3: The Bureau of Labor Statistics should continue to study and, where possible, obtain information on secondary activities that are to be covered in satellite nonmarket accounts.

There are other potential concerns about the ATUS that the panel would consider secondary to those described above. One is that the ATUS obtains information only from one person in each household. While most of its American predecessors were no different, the 1975-1976 time-use survey conducted at the University of Michigan did obtain diaries from both spouses (in married-person households), and that practice is increasingly common in other countries (e.g., recent surveys in Australia, Germany, and Korea). For some kinds of academic research—for example, work examining household bargaining and marital sorting—having diaries from both spouses is crucial, since idiosyncratic spousal interactions are central to understanding this behavior. For purposes of measuring how a population allocates its time, however, the fact that the ATUS does not collect data from multiple family members seems less important. In constructing nonmarket accounts, we are interested primarily in how much time individuals in different demographic groups and with different economic characteristics allocate to the various activities whose valuation might be included in the accounts. Spousal interactions may be important, as spousal complementarities probably generate substantial additional value, but including the value of this intrahousehold capital is a more difficult and subtler matter than the central one of obtaining good estimates of how people spend their time.

With higher levels of funding, it might be desirable to attempt to obtain time budgets from several (or even all) household members. One reason for this is that, as it now stands, ATUS cannot be used to calculate how much combined time parents spend with children in two-adult households. The total time spent with a child depends on whether mom's time and dad's time are positively or negatively correlated, and ATUS provides no information about this. This would be a problem, for example, for a "family care and human capital" account. Whether such a redesign would be feasible within the existing ATUS framework is an open question; Canadian experience suggests that the fraction of households from which multiple household responses could be collected for a single day would be significantly lower than the fraction from which a single response could be collected. In any event, in the presence of limited resources, it is the panel's judgment that this is not a top priority. It would be useful to conduct some split-sample experiments comparing the alternatives in each case—data collected from multiple adults in the household versus a single adult, and data collected for multiple days versus a single day—in order to have a firmer basis for assessing the benefits that might be associated with the multiple adult or multiple day approaches.

A second "minor" concern is that the ATUS asks each respondent for a time budget for only one day. Only the 1975-1976 U.S. survey obtained diaries on multiple (four) days, although the leading international time-budget surveys have typically obtained data for two days. Having time budgets for multiple days would allow researchers to address a variety of interesting issues, particularly those related to variation in the timing of activities. Obtaining diaries on multiple days also carries the potential to reduce sampling costs. On the other hand, because respondents are drawn from the CPS outgoing rotations, any potential advantage with respect to sampling costs may be less for the ATUS than for a time-budget survey that requires a fresh sample. Perhaps more importantly, asking respondents to report for multiple days could have the serious disadvantage of depressing survey response rates still further. In the end, the relative value of having multiple reports from particular respondents as opposed to single reports from a larger number of respondents may depend on whether interday or interpersonal heterogeneity in time allocation is greater. For purposes of constructing nonmarket accounts, the gains from having data on time use for multiple days seem quite small. It is true that the activities reported by any particular individual on any particular day may be unrepresentative of how they typically spend their time, but such anomalies should average out across the population of respondents.

Another limitation of the ATUS from the nonmarket accounting perspective is that data are collected only for people aged 15 and older. The exclusion of children and young teens means that other data will be needed to quantify the time spent in school or school-related pursuits, as would be required to construct an education satellite account.

A final point here is that different sorts of time-use data may be needed to examine the tradeoffs that households face when weighing unpaid production and market substitute options. Learning about these tradeoffs would require information about how individual households combine time with purchased goods and services to produce the various things they need and enjoy in daily life. Analyses based on ATUS data might create a few demographic cells (e.g., age and education), construct estimates of average time expenditures for those cells, and link them to average goods expenditures from a consumer expenditure survey. Such analyses could be informative with regard to how individuals, on average, combine goods and time (see, e.g., Gronau and Hamermesh, 2003), but they would miss all of the idiosyncrasies inherent in marital matching and household behavior. It might be valuable to have a one-time survey, perhaps for a sample of households that had previously responded to the Consumer Expenditure Survey, that would provide time budgets and consumer expenditure information for the same people.

The ATUS is not perfect for purposes of constructing nonmarket accounts: it could not be, given budget constraints and the conceptual and measurement difficulties inherent in obtaining time-budget data. We understand that there were good operational reasons for the decisions made in designing the ATUS. There was evidence, for example, that, had the survey been designed to collect time-use information from multiple members of responding household on a particular day, survey response rates would have been much lower. Similarly, testing carried out during the development period raised serious concern that probing systematically for secondary activities in which respondents might have been engaged would have greatly increased the perceived survey response burden and thus adversely affected response rates. And BLS is well aware of the potential for nonresponse bias and has planned research to assess its significance. Still, as work proceeds on the ATUS and on time-use data collection more generally, the limitations and potential biases in the data currently being collected for nonmarket accounting purposes should be kept in mind, and efforts to improve the data pursued.

The criticisms of this section notwithstanding, the ATUS is a tremendous step forward for the federal statistical system. Indeed, without something like the ATUS, one could not seriously contemplate the creation of nonmarket accounts for the United States.

DEMOGRAPHIC DATA

Time-use and demographic data must be combined to provide a firm foundation for nonmarket accounts. Time-use data can be used to answer questions about what individuals with given characteristics are doing with their time; demographic data describe the distribution of these individual characteristics in the population. Time use varies significantly across population subgroups. For example, in general, individuals with young children have less time for certain activities (e.g., traveling, work, going out at night) than adults without children. In addition, the value of time spent may vary with an individual's characteristics. A higher value may be placed on time spent completing 4 years of college than on time spent completing 4 years of high school, for example, because of the greater value of forgone earnings for someone who already has completed high school. Detailed demographic data are needed to estimate differences in time allocation patterns across various socioeconomic subgroups of the population.

There are several reasons that a comprehensive demographic database is not available for the United States. First, in our decentralized statistical system, agencies commonly specialize in producing certain types of data, and these efforts typically are not coordinated. For example, the National Center for Education

Statistics is a source of information on preschool and early grade enrollment by mother's education and employment status, and the National Center for Health Statistics is a source of information on health care visits by age, sex, and race. Second, statistical and methodological revisions frequently are not carried back through time. For example, the Current Population Survey used the 1980 census population controls from 1981 through 1993 before switching to 1990 census population controls in October 1994. This procedure may significantly affect estimates of the absolute numbers of school enrollees and the comparability of the enrollment series across time; other measures, such as enrollment rates, may not be affected (U.S. Census Bureau, 2002, p. 131). Additionally, the available information may not be cross-classified by all dimensions of interest. An attempt is made to minimize respondent burden and, accordingly, a given survey may collect information on only a few demographic variables. A project that attempts to link administrative record information contained in establishment surveys and household surveys is being undertaken at the Census Bureau (Abowd et al., 2004); such projects may facilitate the creation of a unified set of demographic statistics. Although we recognize that creation of better coordinated demographic data would require significant effort by statistical agencies and other suppliers of data, the development of such data is an important goal.

> **Recommendation 2.4:** A consistent and regularly updated demographic database should be assembled as an input to nonmarket accounts. The database should include information on the population by age, sex, school enrollment, years of education and degrees completed, occupation, household structure, immigrant status, employment status and, possibly, other dimensions.

It would also be ideal to have information on health in this demographic database, but this may not be a realistic goal. Household structure data should include demographic information on members of a household, including relationships if any (e.g., divorced or married), and numbers and ages of children and unrelated individuals, as well as information on children not living within the household.

The fact that high-quality demographic data already are collected by the Census Bureau and other agencies makes this recommendation feasible to implement. Data currently collected by the decennial census and by the Current Population Survey provide much of the necessary input. What is needed is to have key data sources linked together, made consistent over time, and published at intervals to support accounting efforts effectively. Many critical census projections are made annually, but these do not include all of the variables that are needed. Most of the essential demographic data fields are included in the census long form but in the past some have been presented only once every 10 years. The Census Bureau's new American Community Survey will (contingent on funding) be continuous, though rolling geographically, and will advance the effort to produce a more fluid demographic description of the population. Given the budget constraints of the statistical agencies, the demographic database should be built, to the maximum extent possible, using existing or already planned information, minimizing the extent of any new data collection.

OTHER DATA NEEDS

In addition to the above-described data, which relate mainly to labor inputs, a complete nonmarket account must include values of nonlabor inputs. For example, a home production account must include data on the capital services, materials, and energy inputs that complement unpaid labor in generating home-produced outputs. Purchases of materials used in home production already are included in the NIPAs, as consumer goods on the production side and as returns to capital, labor, and other inputs on the income side. The NIPAs also include spending on consumer durables, such as refrigerators and washing machines, though the annual flow of services associated with the stock of consumer durables does not correspond on a year-by-year basis with spending on purchases of consumer durables in the same year (see Fraumeni and Okubo, 2001). In accounting for household production, it is the flow of services from these durables that is relevant and for which data are required.

Full development of nonmarket accounts also will require further research and data development to advance understanding of age-old questions relating to the definition and measurement of output. What are the outputs of the various nonmarket activities? Zvi Griliches observed that "in many service sectors it is not exactly clear what is being transacted, what is the output, and what services correspond to the payments made to their providers" (Griliches, 1992, p. 7). This observation is especially pertinent for many of the areas of interest here that are dominated by services—and services difficult to measure at that—such as education, health, social services, culture and the arts, and recreation.

The need for development of better measures of nonmarket outputs can be illustrated with reference to education and health. Frequently, in difficult-to-measure sectors, the value of output is set equal to the aggregate value of the inputs used in its production. Accordingly, little is known about growth, quality improvements, or productivity in these sectors. In recent years, alternative approaches have been developed for estimating educational output more directly. Examples of these approaches include indicator (e.g., test-score based) approaches, incremental earnings approaches, and housing value approaches. Similarly, for a health account, data on the population's health status, of the sort now being developed in disease state and health impairment research, hold promise of providing direct measures of the output of the health sector. The data that will be needed to create these output measures, as well as the data required to construct defensible measures of other sorts of nonmarket production, are discussed at the appropriate points in Chapters 3 through 8.

3

Home Production

Early national income accountants acknowledged that household services represented productive work. A landmark National Bureau of Economic Research study of income in the United States (King, 1923) calculated the value of household services based on estimates of the number of women age 16 and over primarily engaged in housework without monetary remuneration. Assuming that the proportion of "housewives" to the total population of women not employed for pay remained constant and that the average value of their services in 1909 was about equal to the average income of persons engaged in the paid occupation of "domestic and personal service," the value of housewives' services amounted to 30.7 percent of market national income in 1909 and 25 percent in 1918.

Despite this initial interest in home production, national income accounting moved in a different direction. Pigou (1920, p. 11) insisted that national income should be defined only in terms of goods and services that could be brought "directly or indirectly into relation with the measuring rod of money" and discouraged the application of such a measuring rod to household work. A national income estimate prepared in 1926 under the direction of Francis Walker, chief economist of the Federal Trade Commission, explicitly excluded the value of housewives' services (Carson, 1975). In 1929 Simon Kuznets joined the National Bureau of Economic Research and continued the work that eventually formed the underpinnings of the National Income and Product Accounts (NIPAs) that we know today. He conformed to immediate precedent, declining to include any imputations of the value of household production, but offered eloquent warning of the limitations of the resulting estimates of national income: "The welfare of a nation can scarcely be inferred from a measurement of national income as defined [by the GDP]" (Kuznets, 1934, report to Congress).

More recently, there has been a renewal of interest in the development of satellite household production accounts. A number of individual scholars have made serious efforts to lay out a framework for such accounts (see, e.g., Ironmonger, 1996; Landefeld and McCulla, 2000). Several national statistical agencies are in the early stages of developing satellite household production accounts. Among other efforts, Eurostat is funding several country-specific projects, and the Office for National Statistics in the United Kingdom has developed an account that estimates a value for that country's household production (see Holloway et al., 2002). A major catalyst for this activity—and for prospects of future progress—is the development of time-use surveys in a number of countries.

In this chapter we analyze the nature of household production and discuss the issues involved in measuring its contributions to an expanded notion of gross domestic product (GDP) and to the construction of satellite national accounts. In keeping with the purpose of developing a framework for satellite accounts, we consider how to measure the quantities of the various inputs used by households to generate their products and how to assign values to those inputs. We then turn to issues of classifying and valuing the outputs from home production, which we argue should be done independently from the measurement and valuation of input quantities so as to maintain the double-entry bookkeeping that is the foundation of all accounting.

Accounting for the value of what is produced in households always has been conceptually necessary for obtaining a complete picture of what society produces. Indeed, in a prehistoric or undeveloped society, most of what was produced would have been outside the market; the only sensible "national" accounts would consist almost entirely of household production. With the growth of markets and of trading states and the subsequent development of expanded trade in manufactured goods and then in services, the relative importance of home production in total final output clearly diminished. Researchers, typically working with data from European countries, Australia, or Canada, have produced varying estimates of the magnitude of output from the household sector. Niemi and Hamunen (1999) estimate that the value of household output ranges from about 35 to 55 percent of GDP.[1] All of these estimates are rough since neither the physical magnitude of the output nor its value have been precisely measured.

Basic demographic information allows another way to get a rough indication of the amount of household output relative to the market output of the nation About two-thirds of the U.S. population over age 16 is in the labor force (U.S.

[1]Some of the differences in findings across studies reflect definitional differences, in addition to real differences in the importance of home production across countries. For example, some studies include the value of volunteer labor to organizations as a component of home production; others do not. Likewise, some studies attempt to net out certain components of home production, such as housing services, so that there is no double counting with GDP.

Census Bureau, 1999). If we assume that an adult spends 10 hours each day in sleep and basic personal care, then those who are in the labor force, if they devote 40 hours each week to "work," expend only about 40 percent of their waking hours "at work." The other one-third of the population spends none of its time in the labor market producing output that is captured by GDP. So, overall, the adult population spends only about 25 percent of its waking hours "at work." Depending on what one chooses to define as productive output in the nonmarket sector, how productive those nonmarket hours are in comparison with those spent "at work," and how one values the resulting output, the nonmarket product could easily be greater than the whole of GDP! We do not make the case so strongly, but we note the potential magnitude of this nonmarket product and stress that the difficulty in measuring it does not imply that it is not important in magnitude or in policy significance. Indeed, we argue that, in judging the level and composition of national output, anyone who disregards nonmarket product misses a significant part of the picture.

Estimates of the value of household production are relevant not only to measures of the level of economic activity and productivity, but also to their trend rates of growth and to their fluctuations over the business cycle. Many historical estimates of growth in the output of the U.S. economy during the nineteenth and early twentieth centuries were constructed rather crudely, using information on relative rates of growth in employment in different sectors. When the output of the nonmarket household sector of the economy is imputed using a similar methodology, the estimated trend growth rate of overall output may be substantially modified (Folbre and Wagman, 1993). Time and energy devoted to nonmarket work vary less over the business cycle than the corresponding inputs to market work, and may even move countercyclically as unemployed workers choose to devote more time to home-based activities (Greenwood et al., 1995). Thus, if household production is omitted from measures of total output, it may lead to misleading conclusions about patterns of economic growth.

The productivity of nonmarket work is modified by technological changes, such as the growing availability of consumer durables. The development of microwave ovens, for example, reduced the time necessary to produce meals at home. Increases in the efficiency of home meal production need not mean that time devoted to home meal preparation will decline—increases in efficiency may encourage more meal preparation at home than would otherwise have taken place. More recently, home computers and the flourishing of the Internet now allow households to perform a range of tasks—pay bills online, communicate by e-mail, shop on the Web, produce photos for distant relatives—either more quickly or better. While the productivity of time devoted to household work has risen, wages—the returns to market work—also have grown, possibly even faster. All else the same, relatively more rapid growth in market as compared to household productivity implies a price effect that would lead to substitution of time toward market production.

A dramatic recent historical trend related to household production has been the increase in women's participation in wage employment—a shift of women's efforts from the unmeasured realm of nonmarket output to activities that are valued in GDP. Accounting for this shift and the resulting reduction in the time available for nonmarket work would change measures of the growth of aggregate output over time (see Goldin, 1986, for documentation of these trends). From a cross-sectional perspective, taking the value of nonmarket work into account in calculating household income might reduce measured inequality. The valuation of household time also has implications for the construction of poverty rates and other measures of relative welfare.

Bringing household activities into satellite accounts raises significant classification problems. Measuring quantities produced, as opposed to consumed, is one challenge. For example, expenditures on food at McDonald's are included in GDP, the cost of which includes the wages paid to McDonald's employees; however, the value of time customers spend eating McDonald's meals is not—and should not—be included because we are interested in measuring the production of output that may contribute to consumers' welfare, rather than in measuring welfare itself. An analogous distinction needs to be made when attempting to measure the quantities of output produced at home. Deriving meaningful valuations of nonmarket inputs is another challenge. The value of market inputs is measured by their prices, which presumably reflect society's valuation of them at the margin. By their nature, however, many of the inputs into household production, particularly the time household members devote to such production, are not purchased. The inherently nonmarket nature of home production means that the value of these inputs must somehow be imputed rather than measured directly.

Beyond the distinction between household production and household consumption, productive activity may be defined more or less broadly. This chapter focuses on household production of tangible goods and services produced for immediate consumption, together with household production of tangible investment goods, such as the building of a home addition. At one level, care of children should be considered a component of home production. If parents wished to do so, they could hire someone else to change their infants' diapers, make their children's meals or help their children with their homework, so the production of these services could be assigned a market value. The services of hired child care providers may be, at least beyond some point, a highly imperfect substitute for parental time devoted to the same set of tasks. There is, moreover, an investment dimension to the care of children that the valuation methods advocated in this chapter simply do not capture. The issues surrounding family care, especially family care of children, are discussed at greater length in Chapter 4. Similarly, although the time-use estimates reported later in the chapter include time devoted to educational activities, the methods we advocate for measuring home production of tangible goods and services are not applicable to measuring the production of human capital assets. Time devoted to educational activities and the production of human capital should be accounted for in a separate satellite.

On balance, the availability of relevant data and the relative simplicity of underlying concepts make the household production account a leading candidate for immediate development. Despite the difficulties outlined here, the theoretical and empirical bases for the inclusion of household production in satellite national accounts on a continuing basis are surely far sounder than was the understanding of accounting for market transactions when the United States made the leap into creating consistent sets of national accounts beginning with data from the 1920s and 1930s.

Recommendation 3.1: The Bureau of Economic Analysis should create satellite accounts reflecting the quantity and value of inputs and outputs in home production.

THE HOUSEHOLD AS A FACTORY

Individuals do not simply consume purchased goods: they combine these goods with inputs of their own time to generate what Becker (1965) referred to as commodities. Thus, a commodity, "lodging" for example, is produced using a purchased (or rented) housing stock and family members' inputs of time into its maintenance and improvement. Meals are produced mainly at home, using purchased materials—raw food together with the services of kitchen appliances, tableware, napkins, and so on—and also the time of those who prepare the meals and clean up afterwards; inputs and output for this simple example are listed in Table 3-1.

As with any good that is produced in the market, there are a variety of ways of generating a given amount of any home-produced commodity. One could generate meals entirely through market purchases, by employing cooks who prepare meals using the purchased inputs noted above and cleaning persons to clear and wash the dishes. At the other extreme, one could grow one's own vegetables, feed and slaughter one's own animals, and use one's own time to prepare the meals and to clean up afterwards. As another example, one could generate lodging by buying a house, then employing a butler to coordinate and effect all repairs. Alternatively, one could build the house and do all repairs and maintenance oneself, perhaps using some purchased materials. In each case, there is a large range of choices between the extremes. As with a market-based production technology, the household production function indicates for each commodity the tradeoffs between efficient uses of different methods to generate any given amount of the commodity.

The choices that individuals within a household make about how to produce commodities depend on their preferences, the relative prices of the goods and household members' time that might be used to generate the commodities, their

TABLE 3-1 Stylized Account for Home-Produced Meals in a Household Production Account

Inputs	Output
Family members' time	Home-cooked meal
Food shopping	
Meal preparation	
Meal clean-up	
Purchased materials	
Food	
Dishwashing detergent, etc.	
Services from consumer durables	
Stove	
Refrigerator	
Microwave	
Dishwasher	
Housing (perhaps)	

comparative advantages (which are linked to personal skills and abilities), and their production technologies. Otherwise identical households in which household members have higher opportunity costs of time, typically reflected in higher market wages, will be less likely to produce commodities at home and more likely to purchase market substitutes. Transaction costs contribute to cross-household variation as well. People who live far from towns incur higher transaction costs to eat in restaurants and should, all else equal, be more likely to eat at home than those who live in town. Similarly, the transaction costs of going out for breakfast or a morning cup of coffee may be high for someone who is typically at home in the mornings, and comparatively low for lunch if the person works during the day at a location convenient to restaurants. Other characteristics of a household, such as its size, may affect the choice between home production and market alternatives. Continuing with the meals example, larger families will benefit from economies of scale in converting time spent shopping and preparing meals, together with needed ingredients, to produce home-cooked meals.

The relative bargaining power of individual members over household decisions also may influence time allocation. When married mothers increase their hours of market work, husbands may not increase their hours of nonmarket work commensurately. Widespread awareness of this pattern has contributed to a proliferation of studies of how men and women bargain over the allocation of time and responsibility, as well as money, in the household (Lundberg and Pollak, 1996).

A household's preferences and its choice of technology generate the combination of home-produced and market-purchased commodities that make it as well off as possible. The choice of household technology can be seen in Figure 3-1,

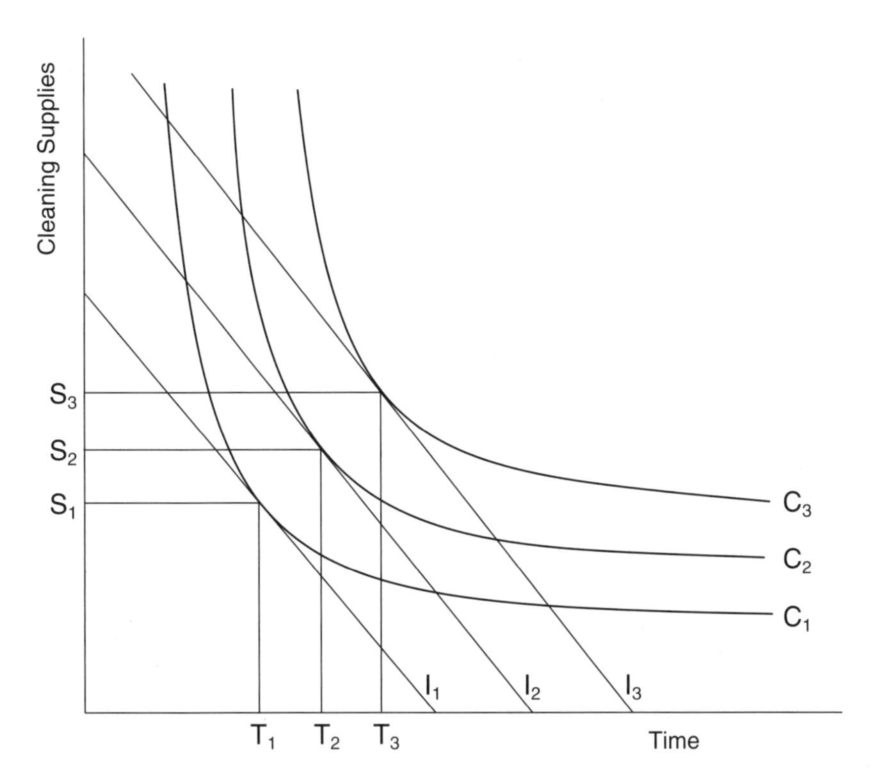

FIGURE 3-1 Production function for home cleanliness.

which shows a household's production of the commodity home cleanliness, obtained through the combination of adults' time and purchased goods, summarized as cleaning supplies. Various outputs of cleanliness, denoted by isoquants C_1, C_2, and C_3, can be produced, with each amount generated by various combinations of cleaning supplies and adult time. Increases in either input raise the amount of cleaning produced. The lines I_1, I_2, and I_3 indicate the relative prices of cleaning supplies and adults' time, with the lines being flatter if cleaning supplies are relatively more expensive, and steeper if the opportunity cost of the adults' time is higher. Each tangency of these price lines with an isoquant curve shows the best possible combination of cleaning supplies and time to produce that output of cleanliness.

Decisionmakers may derive pleasure from having cleaner homes, but they must weigh the extra pleasure against the fact that, in generating a cleaner house, they reduce the time available for the production of other commodities that they enjoy consuming, for consuming directly in the market, or for leisure, which is also enjoyable. If they produce C_1, they leave a lot of their resources for the production of other commodities; if they produce C_3, they leave very little. They will produce that amount, perhaps C_2, which is consistent with their preferences between home cleanliness and all other commodities they might produce. The household will choose to produce/consume an amount of cleanliness indicated by C_2, using inputs of cleaning supplies S_2 and adult time T_2. Clearly, as cleaning supplies become less expensive or adult time more valuable, home cleaning will be produced with more cleaning supplies and less adult time. Of course, there are some constraints on the extent to which such tradeoffs can be exploited; some minimum levels of time and supplies are needed to do the job, as indicated by the nearly vertical and horizontal portions of the isoquants. Whether the home becomes more or less clean (whether the household moves outward or inward from C_2) as household members' value of time increases depends on how much time cleaning the house takes relative to other household-produced commodities and on household members' preferences for cleanliness versus other home-produced or purchased commodities.

Economic theory posits that efficiency is served, or waste is minimized, when people allocate their time to its most valuable use. That use depends on the technology at home and in the market, their skills in each, and the value placed on the products they get from these alternative ways of using their time. As productivity in the labor market rises, through their (or their firm's) efforts, they have greater incentive to spend more time in the labor market. This implies that, all else equal, as society becomes richer due to higher market labor productivity (and thus a higher value of time), people who work for pay will substitute away from home production, which is relatively time intensive, and toward market production. There may be a counterbalancing effect on market labor supply in developed economies if greater longevity and higher lifetime incomes translate into a relatively higher fraction of the population that has retired from market production.[2] Still, we would expect home production performed in high-income societies to be relatively goods-intensive—time spent in home production will be combined with ever-increasing amounts of market-purchased goods.

The key point in this discussion is that economic growth can alter the relative importance of home and market production. That in turn may lead to incorrect inferences about how fast average economic well-being is growing if only market GDP is measured. Since changes in the scope of home production will differ across income groups as their opportunities change, ignoring it in measuring incomes also will bias conclusions about how inequality is changing.

[2]There is some quantitative evidence that as household members transition into retirement, expenditure on food declines sharply while the amount of time spent on home food production increases. If these effects offset one another, and neither the quantity nor the quality of food intake deteriorates, "marginal value of wealth [their term] remains unchanged, even though the market component of it has declined" (Aguiar and Hurst, 2004).

MEASURING INPUTS

The first step toward measuring inputs to household production—especially the time devoted to household production—is to define what is meant by home production, distinguishing productive activities from those that must be viewed as purely consumption. As with any accounting scheme, the decisions are necessarily arbitrary. At one extreme, one could argue, for example, that the sex act is an investment activity—in a few rare instances it generates children who can be viewed as yielding future streams of services, and it may more frequently enhance the participants' subsequent market and nonmarket productivity. Time spent sleeping, the single largest activity in almost any weekly time budget, might also, at least in part, be viewed as an investment in one's human capital—without sufficient sleep, just as without sufficient investment in skill, one cannot produce. But including these activities as household production (even in the investment-focused human capital account described in Chapter 4) would seem a very long stretch.

At the other extreme, the time devoted to replacing one's roof would justifiably be included as household production. Similarly, time spent taking a sick infant to the pediatrician's office is a good candidate for inclusion in a satellite account for household production (or health).[3] Even these activities, however, contain some consumption aspects: the person who chooses to repair her roof may derive some pleasure from the creative aspects of the job; likewise, the parent who spends time helping a child get well feels substantial satisfaction from that activity. These caveats, however, are no different from problems one finds in measuring the use of time in market activities: although the conventional assumption is that people dislike having to work, many people report that they enjoy their jobs (Juster and Dow, 1985). Yet, nobody objects to valuing their labor-market activity and including the result (their compensation) in the NIPAs. In this sense, paid employment is not particularly different from nonmarket work.

The issue, then, is to decide which home activities are production, and thus merit inclusion in satellite accounts, and which are not. In her discussion of household production, Margaret Reid (1934, pp. 8-10) wrote:

> In separating production and consumption perhaps the first thing which needs to be made clear is that consumption activities are personal. They are activities which the individual carries on to meet his own needs. No person can consume

[3]Home production of goods retained for the household's final consumption and gross capital formation by the household are in fact included within the conceptual production boundary defined by the System of National Accounts 1993 (SNA93), the framework that guides national income accountant design around the world. In practice, most countries' accounts omit even this sort of home production, and the SNA93 explicitly excludes most home production of services for own final consumption (see Organisation for Economic Co-operation and Development, 2002, for further discussion).

> for another . . . if the person to satisfy his own need must carry on the activity, that activity is consumption, not production.

These distinctions between production and consumption, and between substitutable and nonsubstitutable services, are necessarily somewhat arbitrary, relying to some extent on social norms. The third-person criterion nonetheless remains a reasonable guideline, and it would be useful to have a home production account that focuses on activities that follow the Reid distinction.

> **Recommendation 3.2:** A household production satellite account, focused on the production of goods and services for consumption, should include only those household activities that could be readily accomplished using market substitutes for household members' time.

This criterion obviously includes such things as meal production, production of clothing services (e.g., doing the laundry and mending), and production of shelter services (e.g., cleaning and home maintenance). Child care and home schooling might also be included as home production under the third-party criterion, though purchased services may be an especially imperfect substitute for time parents spend with their children, and standard methods of valuing home production are poorly suited to capturing the value of investment activities.

The third-party criterion excludes such activities as sex, sleep, travel, volunteering, studying, and exercise. This does not mean, of course, that these activities are not productive. Studying and exercise, in particular, may be critical inputs to the production of skill capital and health capital, respectively. We consider these activities to be outside the sphere of the (limited scope) home production account contemplated here and discuss them in other chapters. We also stress that home production activities fall across a continuum that spans the Reid criterion and, in some cases judgment is required as to how to categorize an activity.

How should household members' inputs into nonmarket production be measured? Time diaries are the preferred methodology for examining time use (see Chapter 2). The diaries yield responses that are more reliable than answers to stylized questions such as, "How much time did you spend providing child care last week?," which require recall over a longer period and also suffer from considerable variation in personal definitions of child care. Another problem with stylized questions is the temptation for respondents to report what they would like to believe about their own activities or what they think interviewers would like to hear. Selective reporting can occur with time diaries as well but, because of the episodic nature of the reporting, the problem should be less severe.

In Table 3-2 we list activities that seem to meet the criterion for inclusion in a household production satellite account—activities that one could readily envision being performed by market substitutes—and provide estimates of time devoted to them. The time inputs are averages over all individuals (adults only were included) in the 1985 U.S. time-use sample and over all days of the week. They

TABLE 3-2 Time Spent in Home Production, by Sex, Marital Status, and Presence of Children, Americans Aged 18 and Older: 1985, in Minutes per Day

	Married				Single	
	Female		Male			
Activity	Children[a]	No Children	Children	No Children	Female	Male
Household Work						
Food preparation	70.22	68.63	16.09	18.43	43.46	17.57
Meal cleanup	18.48	21.84	3.84	4.32	11.77	3.48
Cleaning house	53.67	42.91	12.82	13.08	31.93	9.03
Outdoor cleaning	3.64	4.40	11.77	18.65	4.98	8.91
Clothes care	24.76	20.36	2.94	1.99	12.92	3.32
Repairs	3.71	4.98	20.00	17.40	2.61	10.74
Plant, pet care	6.01	8.41	8.37	12.71	5.23	3.76
Other household	14.05	15.03	16.80	15.57	12.15	12.50
Total	194.54	186.56	92.63	102.15	125.05	69.31
Child Care						
Baby care	25.98	4.10	3.93	0.78	3.01	0.34
Child care	21.19	4.74	6.30	0.69	6.52	0.69
Helping/teaching	4.22	1.23	1.37	0.06	1.31	0.14
Talking/reading	3.39	0.76	1.08	0.63	1.13	0.29
Indoor playing	8.59	2.69	5.30	1.10	1.43	0.82
Outdoor playing	1.22	0.75	1.11	0.94	0.48	0.16
Medical care, child	1.70	0.40	0.29	0.00	0.24	0.00
Other child care	2.66	2.62	0.70	0.80	1.86	1.10
Travel/child care	12.75	3.19	5.87	1.26	3.10	0.44
Total	81.70	20.48	25.95	6.26	19.08	3.98
Obtaining Goods and Services						
Everyday shopping	31.56	30.89	15.86	17.21	27.69	15.50
Durable/house shopping	1.26	0.66	1.15	0.44	0.62	0.44
Medical appointments	4.27	2.05	2.54	1.78	0.83	0.87
Gov./financial services	1.68	1.83	1.10	1.86	1.54	1.13
Repair services	1.42	0.86	1.69	1.05	1.07	1.83
Other services	2.14	1.89	2.98	1.47	1.90	2.91
Errands	1.99	1.19	2.59	1.22	1.04	1.34
Total	44.32	39.37	27.91	25.03	34.69	24.02
Education						
Students' classes	1.29	1.02	0.32	1.33	12.04	21.78
Other classes	2.02	1.49	1.56	1.51	2.16	3.27
Homework	1.83	3.29	1.72	1.41	9.06	21.86
Other education	0.33	0.13	0.00	0.13	0.50	1.03
Travel/education	8.62	15.86	3.34	3.78	10.52	1.69
Total	14.09	21.79	6.94	8.16	34.28	49.63
Domestic Crafts	8.62	16.86	3.34	3.79	10.53	1.69
Total, hours per day	5.72	4.75	2.61	2.42	3.73	2.48

[a]Under 18 years old.

NOTE: Sample size = 5,358.

thus represent typical activities on an average day in that year by people in the identified demographic groups. The survey from which these estimates are derived, funded by the National Science Foundation and conducted by the University of Maryland, had a sample size of 5,358 respondents, 1,612 of whom were parents. Interviews were conducted by mail, telephone, and in person; response rates for the three modes were 51 percent, 67 percent, and 60 percent, respectively. For comparison, we also include similar statistics based on pooled data from surveys conducted in 1998, 2000, and 2001; see Table 3-3. Though more recent, the sample sizes are smaller and there are no 2000 data for single persons. It should be noted as well that the time-use categories are not fully comparable between the two tables.[4]

Clearly, under the definitions employed in these tables, inputs into household production form a substantial fraction of adults' time, ranging from 10 percent for married men with no children in the home to nearly 25 percent for married mothers with children present. Even excluding child care, household production amounts to more than 2 hours per day for the average American in all six demographic groups. The statistics demonstrate that, independent of marital status or the presence of children, household production, as measured here, is disproportionately done by women. Child care activities may be performed by women more than men at least partly for biological reasons but, even among single individuals without child-care responsibilities, women spend substantially more time on household work and the procurement of goods and services than do men. The observation at the beginning of this chapter that accounting for household production is especially important in measuring total productive activity because of women's changing labor force participation is underscored by the clear dominance of women's time in the set of activities that are candidates for inclusion in a household production satellite account.

As more detailed, extensive and timely data become available, the measurement of time devoted to home production can be fine tuned. Even with more detail, however, measurement problems remain. Perhaps the most serious is that productive activity may occur during periods of time that are categorized as personal consumption or leisure. For example, if a woman lists in her time dairy a 30-minute segment spent watching television, but she was also keeping an eye on her young child's activity, this accounting would ignore what might be included as her time input into household production (caring for her children). Some time-use studies collect information regarding "secondary" as well as "primary" activities. Analyses show that many secondary activities are child-care related.

In cases where both primary and secondary activities are classified as productive, the particular input of time should be counted only once. For example, over the course of a given interval—say an hour—time input should add up to 60 minutes, even if that time was spent in multiple activities. The panel recognizes

[4]For more information on these surveys, see Bianchi et al. (2001).

TABLE 3-3 Time Spent in Home Production, by Sex, Marital Status, and Presence of Children, Americans Aged 18 and Older: 1998, 2000, and 2001, in Minutes per Day

	Married				Single	
	Female		Male			
Activity	Children[a]	No Children	Children	No Children	Female	Male
Household Work						
Food preparation	49.44	40.52	17.94	19.08	25.59	15.68
Meal cleanup	11.47	9.15	3.28	3.82	5.21	1.83
Cleaning house	44.07	31.27	17.40	17.35	19.12	10.99
Outdoor cleaning	6.60	11.00	15.73	28.13	5.39	8.59
Clothes care	28.73	18.12	3.01	4.75	13.61	6.40
Repairs	7.19	4.65	14.77	21.41	4.30	9.68
Plant, pet care	6.13	7.12	3.91	11.48	4.77	3.38
Other household	12.88	19.70	8.58	8.33	7.93	10.82
Total	166.51	141.53	84.62	114.35	85.92	67.37
Child Care						
Baby care	15.60	0.00	4.27	0.00	3.54	0.09
Child care	40.56	3.13	20.89	1.40	11.17	2.28
Helping/teaching	9.55	0.67	3.39	0.28	1.63	1.43
Talking/reading	5.71	5.16	4.49	0.74	3.61	0.43
Indoor playing	9.20	1.00	10.12	2.75	8.39	1.08
Outdoor playing	3.76	0.14	1.97	1.25	1.74	0.00
Medical care, child	1.20	0.00	0.52	0.00	0.69	0.24
Other child care	8.79	0.80	3.86	1.64	3.83	1.08
Travel/child care	14.41	0.55	7.65	1.46	4.54	1.08
Total	108.78	11.45	57.16	9.52	39.14	7.71
Obtaining Goods and Services						
Everyday shopping	9.83	11.15	5.52	3.19	6.78	5.33
Durable/house shopping	22.33	24.82	18.29	10.07	19.15	7.84
Medical appointments	10.34	0.98	1.71	2.47	2.86	1.01
Gov./financial services	1.40	0.88	0.56	0.72	0.70	0.77
Repair services	2.38	0.38	2.67	0.84	0.97	0.47
Other services	0.48	5.90	1.38	0.06	2.27	4.55
Errands	1.17	0.87	0.45	0.31	2.17	1.34
Travel/shopping	25.58	23.58	17.94	14.52	19.28	14.47
Total	73.51	68.56	48.52	32.18	54.18	35.78
Education						
Students' classes	1.38	0.36	1.44	0.26	10.45	14.76
Other classes	3.00	1.10	3.29	1.63	4.24	9.78
Homework	12.13	12.34	18.59	25.04	27.93	32.19
Other education	0.24	0.26	1.36	0.00	0.37	0.00
Travel/education	1.68	0.50	1.23	0.30	3.11	4.35
Total	18.43	14.56	25.91	27.23	46.10	61.08
Domestic Crafts	5.34	10.16	0.98	0.00	1.26	0.08
Total, hours per day	6.21	4.11	3.62	3.05	3.78	2.87

[a]Under 18 years old.
NOTE: Sample size = 2,696.

that single counting a period of time that may be doubly productive (e.g., a primary activity of preparing dinner and a secondary activity of child care) may offend some observers' notions of the contributions of productive household activities. A central principle of national income accounting, however, is to avoid double counting. Suppose, for example, that a mother spends 15 minutes cooking and simultaneously caring for a child. Both the meal produced and the care provided to the child are valuable outputs, but satellite accounts should include only a total of 15 minutes of the mother's time as an input to that home production.

> **Recommendation 3.3:** Quantities of time inputs should be included only once, regardless of how many different productive activities may be accomplished during the particular intervals. If more detailed disaggregion is called for, time devoted simultaneously to multiple productive activities should be divided into parts attributable to each.

If division of time units is desired to account for multitasking, a "value-theoretic" approach might be used. Under this approach, suggested by Nordhaus (2004, p. 19), time in the activities is divided proportionately to their value. So if a parent is monitoring a child and doing laundry for an hour, and the value of the time for monitoring a child is twice that of doing laundry, one would attribute 40 minutes to child care and 20 minutes to laundering. As an alternative to dividing time units more finely, new combined activities—child care and doing laundry—could be defined and valued.

Measuring the quantities of household inputs that generate value added is not easy. It requires obtaining good information on how household members spend their time and making some difficult decisions about which activities are to be counted as production and which are not. Nonetheless, the decisions to be made regarding measurement of time inputs may be the easiest of those required before a set of satellite accounts that includes household production can be constructed.

VALUING INPUTS

Valuation of the time that household members devote to home production raises further conceptual and practical difficulties. The appropriate solutions to these problems, unlike some of those involved in measuring inputs, will not result simply from additional and better quality data. Instead, they will require substantially more conceptual work than has yet been done. Several methods exist for valuing household members' time in production. This section discusses these options, each of which will give a substantially different answer.

The Price of Market Substitutes

By virtue of using the third-party criterion to define what is considered to be home production, it is close to a tautology to say that market substitutes generally are available for the resulting outputs. The wage rates of the workers who contribute the labor component of these market services provide one basis for attaching a price to the time used in household production. Under a specialist wage approach, for example, the productive time spent repairing one's roof would be valued at the hourly wages paid to roofers; the productive time spent gardening would be measured at a gardener's wage; the productive time spent changing diapers would be measured at a nanny's hourly salary; and the productive time input into doing laundry would be valued at the hourly wage of a laundry service worker hired to work in the home. Under a generalist wage approach, the wage for a handyman or a domestic servant might be used to value time devoted to a variety of activities.

The logic of using the market approach is clear: it reflects the hourly value of skills applied to particular activities. As such, it provides a market measurement of the contribution of the activity to aggregate production, independent of the consumption and production choices made by the household. The difficulty with both the specialist and generalist approaches is that they ignore factors relating to the efficiency and quality of time that individuals spend in the productive task in the household. This issue is not much of a problem if the productive activity in question is making toast; if it is plumbing, however, an inexperienced homeowner would spend far more time on the task than would an experienced specialist. Using the specialist wage rate thus would overstate the value of the input.

Household Members' Wage Rates

Another approach is to value each household member's time at his or her market wage rate. The argument is that he or she is presumably foregoing some earnings in order to produce at home, so that the opportunity cost of the home production is the foregone earnings. These foregone earnings may be proxied by the person's average wage rate. One benefit of this approach is that it is easy to apply. There is no need to decide on specific market substitutes for each different household production activity and to measure their prices. All quantities of productive household time can be summed and valued at the wage rates of the household members engaged in home production. Indeed, the aggregate amount of time spent in what have been determined to be household production activities could be valued at something close to the national average wage rate.[5]

One of the several problems with this approach is that some home production is generated by individuals who do not have a market wage—housewives and househusbands, retirees, some teenagers, and others. This difficulty can be overcome by imputing market wage rates of such individuals as equal to those of

[5]Using the national average wage rate would not be precisely correct, since those with high market wage rates likely spend relatively few hours in home production, while those with low market wage rates likely account for a proportionally larger share of home production hours.

otherwise identical workers, adjusted using now-standard techniques (Heckman, 1976) to account for the selection of workers with relatively more attractive labor-market opportunities into paid employment.

A significant difficulty with using the household producer's market wage rate is that it fails to reflect the value of time spent in specific, and varied, home production activities. Using a professor's hourly salary to value his time in grilling steak, for example, imputes a high price to an activity that is fairly low skilled. Similarly, valuing the time that any chief executive of a large corporation spends in child care by the wage associated with that position surely would overstate the resulting contributions to the output of the household sector. Even if the household activity is similar to that which an individual performs in the market, absent the pressure to meet production goals and possibly the capital stock in the home to accomplish the task efficiently, that individual may require more time at home performing the activity. Moreover, as discussed in Chapter 1, there is the further complication that an individual's average wage rate may be a poor proxy for the opportunity cost of an hour devoted at the margin to nonmarket production.

An Alternative Approach—Quality-Adjusted Replacement Cost

It should be clear that neither the replacement cost method nor the use of household members' market wage rates are fully satisfactory for valuing time inputs to home production in a satellite account. An alternative approach that addresses the problems with both the replacement cost method and the own market wage method values the time of the individual performing the activity at his or her productivity-adjusted replacement wage, $\hat{W}_R$. As described in Chapter 1, this modified replacement method also establishes a logical bridge between the standard replacement cost and the opportunity cost approaches. This measure of value is

$$\hat{W}_R = bW_S, \tag{3.1}$$

where W_S is the specialist wage for the task and b is a number, typically between zero and one, that indicates the shortfall (or in rare cases excess) of the household member's productivity in comparison with the specialist's productivity in the activity. In the case of toast-making, b is probably close to one. In the case of plumbing, b may be very small, so that $\hat{W}_R$ is nearly zero. (In the case of the professor who, while fixing his toilet, got the snake caught and had to call a plumber, $\hat{W}_R$ clearly was no bigger than zero and arguably negative!) We believe that such a modified replacement approach represents the conceptually ideal method for valuing time inputs to nonmarket production.

Recommendation 3.4: Time inputs to home production should be valued at their replacement cost, ideally adjusted to reflect skill and effort differences between home and market production.

Implementing this recommendation will require that methods be developed to estimate the productivity of the typical household member in different activities and to adjust replacement wage rates accordingly. Specifically, one would like to be able to identify when the skill and effort of people performing nonmarket tasks diverges sharply from the market-based alternatives. Nonmarket accounting of home production—as well as volunteer labor, health care, and other activities—would benefit from research and data that allow estimation of the relative efficiency of nonmarket and market labor. This would require details on the amount (and quality) of work performed in nonmarket environments that goes beyond what is captured in the American Time Use Survey (ATUS).

In thinking about the various possible approaches to the valuation of time devoted to home production, it is interesting to note that individuals frequently undertake household production tasks in situations where their market wage exceeds the amount they would need to pay someone else to do the job. This means that the market value of the individual's time input is less than the amount he or she could have earned by devoting the time to market work. We interpret the difference between the family member's market wage and the replacement cost of the market-based service as an estimate of the consumption value (enjoyment) that the individual receives from supplying the service personally rather than through the market. We do not recommend that this consumption value be included as output in the nonmarket household production account. There are a whole range of activities, including market work, that provide different levels of satisfaction across individuals.[6] As is discussed elsewhere in this report, one could envision a separate account designed to quantify and value the recreational component of activities such as jogging or woodworking that might also appear in other accounts (e.g., health or home production). The goal of the household production account we are proposing is to quantify and value market-replaceable goods and services produced by the household for consumption. Satisfaction derived in the process does not belong in this account.

---

[6]In fact, surveys of individuals seem to indicate that among work and nonwork activities, market work (people's jobs) tends to be in the middle in terms of enjoyment. The cluster of disliked activities includes things like house cleaning, laundry, and going to the dentist (see Nordhaus, 2004, p. 17).

An Example

A brief illustration may help to clarify the distinctions among the concepts described above. Any one of the activities listed in Table 3-2 could serve as our example, but a simple one is outdoor cleaning—chiefly yard work, lawn mowing, and so on. Assume that the time devoted to this activity on a typical day in 2002 was the same as listed in Table 3-2 for 1985. In 2002, the average earnings of groundskeepers and gardeners (the occupation that probably provides the closest market substitute for the time that the household provides) was $10.48 per hour.[7] The weighted (by the relative sizes of the six demographic groups shown in Table 3-2) average earnings of the population of adults in 2002 was $20.87, practically double the average gardener's wage. Multiplying the weighted average of daily minutes of outdoor cleaning by (365/60) yields an estimate of annual hours of outdoor cleaning per adult. Multiplying this by the number of adults in the United States in 2002, roughly 214 million, yields an estimate of the total time spent by adult household members in the United States in 2002 on outdoor cleaning, 10.78 billion hours.[8]

If one values this time input at the cost of a gardener's time (the specialist approach), the implied value of household production in this single category was $113 billion in 2002. Similar calculations could be made using the replacement wage rate for a generalist (e.g., housekeeper) occupational category. Using the average wage rate method (the opportunity cost approach), the implied value of household production in this single category was $225 billion in 2002. To implement our preferred method, the quality-adjusted replacement cost approach, one would need to know how efficient the average household member is in performing the chores that he or she includes under the rubric "outdoor cleaning" in a time diary as compared to a hired gardener. It is unlikely that the average household member is twice as efficient as the typical hired gardener. Indeed, it is more likely that the average household member is less efficient than the typical gardener, so that b in equation 3.1 is positive but less than one, and the person's time input should be valued at less than that of the gardener. This approach implies a value of time devoted to outdoor cleaning of something less than $113 billion. Depending on the method used to value household time in home production, a set of satellite accounts would value home inputs of labor in outdoor cleaning in the United States in 2002 at some positive amount that is less than $225 billion.

It is worth pointing out also that variation in skill and effort on the input side can translate into variation on the output side. The quality of home-produced output may differ predictably from the alternative produced commercially. A

---

[7]This and the other calculations are based on information in the 2002 outgoing rotation groups of the Current Population Survey (http://www.bls.census.gov/cps/cpsmain.htm).

[8]Such calculations illustrate why a demographics account is needed to supplement the satellite nonmarket accounts.

restaurant meal, for example, might be of higher or lower quality than one produced at home and, in principle, this should be taken into account.

Valuing Nonlabor Inputs

While the time of household members constitutes the major input into household production, purchases of materials also are important, as are the services provided by capital inputs.[9] Purchases of materials used in home production already are included in the NIPAs, as consumer goods on the production side and as returns to capital, labor, and other inputs on the income side. The NIPAs also include spending on consumer durables, such as refrigerators and washing machines, though the annual flow of services associated with the stock of consumer durables need not correspond on a year-by-year basis with spending on purchases of consumer durables in the same year (Fraumeni and Okubo, 2001). In accounting for household production, it is the flow of services from these durables that is relevant. For housing, the NIPAs incorporate an imputed measure of the flow of housing services, which is conceptually also appropriate for a household production satellite account.

If household production is ever integrated into the NIPAs, it would be important not to double count the contributions of purchased materials or the flows of housing services to the output of the household sector. These are counted now as part of final output. In an integrated account, they would be treated as intermediate inputs to (final) home production output. The main effect of bringing the household sector into the NIPA framework would be to increase GDP by an amount equal to the value of the time devoted by household members to home production, although the service flow imputed for consumer durables would also yield a different number from the purchase price used in the NIPAs. For the purpose of providing a complete and unified picture of the home production activities of the household sector, all inputs (market and nonmarket) should be included.

> **Recommendation 3.5:** In addition to labor inputs, satellite accounts describing household production should include on the cost side the value of capital services (including those of consumer durables), other services, materials, and energy used in generating home-produced outputs.

[9]See Ironmonger (1996) for an attempt to include these other inputs, albeit in the context of using opportunity cost to measure the value of all inputs; see also Landefeld and McCulla (2000), who do not use opportunity costs.

MEASURING AND VALUING OUTPUT

As noted above, one of the strengths of national income accounting is that independent estimates of input costs and output valuations provide a useful cross-check on each other. These independent estimates do not play quite the same role in the nonmarket context, where, for the reasons discussed earlier, it need not be the case that, even in concept, the value of the inputs to home production equal the value of the resulting output. We return to this point below, but first address a simpler question, that of distinguishing between classifying commodities produced in the household as consumption or investment. Then we address the issue of generating an independent measure of the output of the household, and lastly return to the relationship between estimated input and output valuations.

The first difficulty in thinking about the output of home production is that the distinction between the main expenditure flows in household production—consumption and investment—is unclear. This is not a matter of distinguishing between time spent consuming and time spent producing. In the discussion above we concluded that only time inputs that are obviously devoted to production—activities for which market substitutes are readily available—should be included in a basic household production account. Rather, the issue here is whether a particular productive household activity contributes to the household's consumption or should instead be treated as an investment activity.

The distinctions between consumption goods and investment goods in the GDP accounts are somewhat arbitrary, but in the interest of simplicity it would make sense for the household production satellite account generally to follow the GDP conventions. This means that the production of housing should be treated as investment, with all other household production outputs treated as consumption goods. One exception might be consumer durables. If we accept that the value of purchased inputs (such as washing machines) should be capitalized, so that the service flow associated with such goods rather than their single period purchase price is registered, then home production of durable items (which would not seem to be a common activity) might be categorized as an investment activity and valued as such. In practice, it is unlikely that an account will be able to separate home time accurately into production of durable and nondurable commodities. The NIPA treatment of different goods and services that are purchased by the household sector provides a good guideline for classifying goods and services produced and implicitly purchased by households on the expenditure side of any satellite account. Thus:

> **Recommendation 3.6:** Household production, and whether to classify activities as consumption or investment, should be treated consistently with their market analogs in the NIPAs. One exception relates to consumer durables, which should be treated as a capital investment, with the value of the resulting service flows treated as an input into home production.

The next important question is what units should be used to measure household production of consumption and investment commodities. The answer parallels the answer to the question of how the quantities and prices that generate value added in home production on the input side should be measured. Once it has been determined that an activity belongs in the home production satellite account, the nature of the good or service produced should be defined in such a way that it has an identifiable analog that is sold in the market. The national income accountant then should obtain market prices for the particular goods and services that have been identified as analogs. At least in concept, measuring the prices of these market analogs should be relatively straightforward, once the difficult problems of identifying the goods and services to be included in the home production account and determining the comparison good have been solved.

Let us return to the laundry example. The input-side calculation tallies the number of hours devoted to laundry and multiplies these by the quality-adjusted replacement cost (or some other measure of the value of the household producers' time). The output-side calculation measures the total quantity of laundry done, estimates what it would cost to have it done commercially, and assigns that value to the household's production of the consumption item, laundry services. This value is assumed to be the product of the services of the household's capital (i.e., the services of a washing machine and dryer), purchased raw materials (i.e., electricity, soap) and labor hours.

Recommendation 3.7: Any satellite account for household production should include an output-based measure that is derived independently of the input-based measure of the value of household production.

As noted above, depending on how they are valued, the cost of inputs devoted to home production will not necessarily equal the value of the resulting outputs. If individuals derive intrinsic satisfaction from engaging in home production activities, they may do things that could have been done more cheaply, relative to foregone earnings, by someone else. Were time inputs to home production to be valued using opportunity costs, the inputs to home production could be assigned a value that far exceeds the value of the outputs, with the difference reflecting the pleasure derived from performing the task in question. This consequence limits the usefulness of opportunity costs for valuing time devoted to home production. Using productivity-adjusted replacement costs instead to value time inputs to home production will tend to push input and output valuations toward one another; because of technology, scale, and transaction cost differences between home and market production, however, they still need not be equal. The extent to which these factors lead to divergent input and output valuation estimates will depend on precisely how productivity adjustment is specified (e.g., average home technology levels could be incorporated into the adjustment and reflected in market and home productivity comparisons).

A great deal of historical evidence suggests that differences and changes in household capital, and also in other aspects of household technology, such as economies of scale, are quite significant. Hence an output-based approach may yield substantially different results from an input-based approach, especially for comparisons across societies or over time. While the statistical offices of Australia and Canada largely rely on valuations based on labor inputs and wages, the statistical office of the United Kingdom has adopted an output-based approach. There is no *a priori* reason for the value of an output-based measure of a household's expenditure on home-produced nonmarket items to equal the value of an input-based measure. Although discrepancies also may exist in the NIPAs for market-produced goods, the source of discrepancies in the nonmarket context is different. In the market case, the two should in principle be equal; that they are not can be attributed to the fact that data sources on the income and product sides differ and to measurement error. In the nonmarket case, there is no conceptual reason that the two valuations should be the same.

By suggesting that satellite accounts for household production include independent output- and input-based measures, each based on measures of quantities and prices, we are recommending an approach that allows a more accurate understanding of the value of home production to be obtained than is possible in systems that rely on one of these two approaches alone. As discussed in Chapter 1, national income accounting methods emphasize the importance of double-entry accounts that estimate the value of inputs and outputs, costs and expenditure. Although we think that measuring quantities and values of inputs and outputs separately is the right way to structure this endeavor, we do not know enough or have the information necessary to implement this principle fully at the present time. Thus, while we believe it is clear conceptually what ought to be done, further research undoubtedly will be needed to make the preferred approach practical.

DATA REQUIREMENTS

Input Quantities and Prices—Time Use

The data appropriate to constructing measures of hours devoted to household production must necessarily come directly from recording people's activities in their homes. The activities listed in Table 3-2, for example, are some of the 87 categories into which all activities were coded in the 1985 time-use survey. One of the many crucial benefits of time-budget surveys is that they force the aggregate of time devoted to all activities to equal 1,440 minutes per day for each person. Until recently, time-diary data for the United States were not routinely available. Fortunately for purposes of constructing satellite accounts during this decade, this situation is changing with the introduction of the American Time Use Survey in January 2003 (discussed in Chapter 2).

Recommendation 3.8: The new American Time Use Survey should form the basis for measuring the labor inputs into household production in any continuing set of satellite accounts measuring household production.

While the ATUS provides the first vehicle for constructing continuing satellite accounts that reflect the quantity of labor inputs used in household production, its existence does not obviate the difficulty of measuring the quality-adjusted replacement cost of these labor inputs and thus of valuing the labor inputs. Ideally, whatever approach is chosen for measuring these costs, it should be adhered to for some years, allowing, of course, for occasional reviews and revisions as the nature of household technology changes. Considerable data on market wage rates for different types of market work are available; the difficulty lies with determining the relative efficiency of home and market producers for each activity in the ATUS that can be viewed as productive. The problem is difficult, but no more so than many others that have been handled satisfactorily in creating and updating the NIPAs. The importance of maintaining comparable definitions over time is one that cuts across all accounts.

Output Quantities and Prices

Problems analogous to those of measuring input prices arise when one attempts to measure the value of output produced in the home, although for output, measuring quantities is more difficult than measuring prices. On the quantity side, one must first select an appropriate market analog to what is produced at home. The more difficult issue is to construct measures of the quantities of the specific items that are produced. Returning to the laundry example above, one needs to determine how many pairs of pants, underwear, etc., are cleaned by the typical household during each accounting period to generate a measure of the market-equivalent output of home production of laundry. In addition, one must recognize that the *quality* of nonmarket products, as well as their quantity, is a big challenge in measurement. Even for a reasonably standardized item, such as a lunch of soup and sandwich, there are many attributes of the product, including the convenience of location and service, the variety of choices, and the freshness of ingredients. Coming up with quantity and quality estimates will require new efforts by statistical agencies; the experience of the United Kingdom suggests that, at least for the quantity question, it can be done successfully.

Having determined the particular quantity analog to a home-produced item, its market price should be obtainable, though new data collection again could be required. Existing sources of information that might be useful include the Consumer Price Index (CPI) research database maintained by the Bureau of Labor Statistics (BLS) and the Personal Consumption Expenditure report, published by the Bureau of Economic Analysis (and which uses the BLS data). Although the CPI is designed to measure price change, not the price level, tens of thousands of prices for individual items are collected every month to support production of the index. Items are selected on the basis of local purchasing patterns, which means that different products and services may be priced in different areas. With some effort, these data could be used to compute average prices for selected specific services that are priced in multiple areas.

Clearly, there are data problems with both input- and output-based approaches to the construction of a satellite account that reflects household production. Again, however, we believe these are no more difficult to surmount than were the problems that faced those who constructed the initial versions of the NIPAs. Moreover, undertaking both would provide an illuminating check on the accuracy of the household production satellite account.

4

The Role of the Family in the Production of Human Capital

The preceding chapter focused on household production of tangible goods and services, mainly items produced for immediate consumption (e.g., meals, clean clothing, basic care of children) but also a limited set of items that represent an investment for the future (e.g., the new roof a homeowner installs on the household dwelling). These tangible items are not, however, the only or even necessarily the most important outputs attributable to the activities of the household. This chapter is concerned with household investments that augment the human capital of household members, especially the investments of time that parents and other family members devote to the care and nurturing of children.

CONCEPTUAL FRAMEWORK

No other outputs are as quintessentially nonmarket as the production and care of children. While parents may purchase assistance with the care of their children, market care cannot fully substitute for their personal attention. Some observers appear to view family care as akin to sunlight, available without cost in effectively unlimited supply and thus of no economic interest. Given recent societal trends, however, it may not in fact be safe to assume the availability of family care. Since the 1960s, there has been increased participation of women in the labor force, higher divorce rates, and growing numbers of children who do not live in traditional nuclear families. To the extent that these trends may jeopardize the availability of the family care taken for granted in the past, they suggest the importance of better understanding how family care affects society's production of human capital. Such understanding would help to inform public policy in a variety of areas, such as the allocation of public spending to family programs. Concern about these trends has created particular anxiety about possible effects on young children and, in turn, a resurgence of interest in how family care, especially inputs of parental time, affects child outcomes (see Ginther and Pollak, 2004, on the effect of family structure on children's educational outcomes).

We do not mean to suggest, of course, that families are important only or even primarily because of the role they play in the development of children's human capital—family life is far richer than any such characterization would imply. For many people, beyond whatever their upbringing may have done to prepare them for a productive life, relationships with family members are a source of deep and continuing joy. It is, however, the contribution families make to their children's productive capacities that makes them of interest in the context of seeking to understand and account for society's nonmarket production.

It would be logical to treat the physical production of children—the 4 million infants who are born in the United States each year—as a component of the human capital produced in the home. If some are inclined to question whether these births represent real investment, they might consider the economic situation in year $t+20$ in the event there were no births in year t! Yet, children are people in their own right, not a commodity to be bought and sold, and many people are uncomfortable with viewing newborn infants as an output to be measured like other outputs. A similar discomfort is manifest, for example, in laws governing adoptions. Many childless couples would be willing to pay a large sum for the opportunity to adopt a child, but laws prohibit their making payments to birth parents in connection with an adoption (other than to cover birth-related expenses).

In the same way that births could be viewed as an investment, it similarly would be logical to treat net immigration to the United States from abroad as an increment to the nation's stock of human capital. Particularly with birth rates at modest levels in comparison with earlier historical periods, immigration accounts for a substantial share of observed growth in population, and changes in immigration rates can have a significant impact on the growth of a country's potential labor supply. Indeed, some countries, such as Canada, have consciously structured their immigration policies to favor high-skilled immigrants who can be expected to make especially significant economic contributions. This is not, however, a theme that we pursue further in this report.

Even taking as given the biological production of children and the raw material with which these children are born, parental time inputs are critical to the development of children's intellectual and emotional readiness to learn. Parental time inputs that create a foundation for learning can be viewed as skill-enhancing investments in human capital. Likewise, family care augments the inputs of the medical care system in the production of health, another investment output that yields a flow of future benefits. It is common practice to treat the skills possessed by those entering the workforce as the product of the educational system, the health of the population to be the product of the medical care system, and so on.

In truth, these outputs are products jointly of the formal institutions in question and the household. Without the foundation and continuing support that is provided by the home environment, observed outcomes would be very different.

Even in nonmarket satellite accounts that focus on the activities of the household, family care is treated not as an investment but rather as an activity that yields output for current consumption. Inputs of care time can be valued at replacement cost, adjusted where necessary and feasible for differences between productivity in the home and the market. Measures of output valuation can be derived from the cost of analogous market services, as is done, for instance, for child care in the home production satellite accounts developed by the Office for National Statistics of the United Kingdom. But this approach is quite distinct from that suggested for satellite accounts to value the outputs of the education and health sectors, in which the production of human capital is treated as an investment yielding a future flow of income or utility.

In this chapter we consider the human capital production function, particularly the contributions that households make to skill capital, health capital, and other forms of human capital. There is a great deal that we do not know about the form of this production function, particularly as it relates to the development of children into well-functioning adults, and we make several recommendations for further research in this area. Given the current state of knowledge, we cannot at present recommend the development of a human capital investment account: not enough is known to proceed in a defensible manner. But it should be recognized that any effort to develop more focused human capital investment accounts—for example, the education accounts or the health accounts that we recommend in later chapters—may yield misleading conclusions if those using the accounts do not recognize the contributions made by prior and concurrent investment activities in the home.

While any comprehensive accounting for investments in human capital seems beyond the range of what is feasible, at least for now, it is possible to say something about the magnitude and the value of home inputs to this investment. The chapter also presents evidence related to parents' investment of time in their children and some thoughts about the valuation of that time. We conclude with a brief description of additional data that would be useful for making progress in this area.

DEFINING HUMAN CAPITAL

One question to be addressed at the outset is what we mean by "human capital." Loosely, human capital investment consists of anything that enhances a person's productive capacities, but this begs the question of which capacities should be considered productive and which should not. Most economic analyses of human capital have focused on market activities. In these analyses, in practice if not in principle, human capital consists of those capacities that raise a person's wage rate, which is presumed to reflect their market productivity. One obvious complication is that not all people work for pay. This problem generally is addressed by imputing a potential market wage to those who are not employed.

To the extent that a person's capacity for home production is not taken into account, such a perspective may be misleading. A child who learns from a parent or other family member how to cook, repair a faucet, or create an attractive garden clearly has acquired valuable productive capacity that ought to be considered a part of their human capital.

Whether within or outside the market, productivity may depend not only on knowing how to do certain things—"hard skills"—but also on other personal attributes necessary to accomplishment—what might be termed "soft skills" (see Carneiro and Heckman, 2003, for an interpretation of evidence on the contribution of family to development of these kinds of skills). Examples of diffuse, hard-to-measure soft skills include the ability to show up on time, to cooperate with others, to show initiative, to learn how to learn, and to cope with stress (Goleman, 1995, offers a popular discussion; see also Duncan and Dunifon, 1998a). Many of these soft skills are related to underlying personality traits that are partly inherited and partly shaped by early childhood experiences. Developmental psychologists have shown that a 3-year-old's ability to exercise sufficient self-control to avoid eating a piece of candy placed on her tongue until given permission to do so is a good predictor of future success in school (see McCabe et al., 2004). Such capabilities are probably correlated to some extent with cognitive skills, but they are not easily inculcated or measured.

The word "skill" may be a misnomer, insofar as much behavior that is economically productive is the result of preferences and habits rather than skills. Motivation and effort are important determinants of success (see Duncan et al., 1996). Individuals who like to work hard will tend to earn more than those who do not, if only because they will work longer hours. Individuals who enjoy contact with others and have so-called "pro-social" or "incentive-enhancing" preferences tend to earn more than those who do not (see Duncan and Dunifon, 1998b; Bowles et al., 2001; Bowles and Gintis, 2000). Not all economically advantageous preferences can be described in such positive terms. Also relevant are willingness to follow orders or to adhere to boring routines, or what behavioral economist William Simon describes as docility (Simon, 1990).

Preferences and habits may have other sorts of effects. A child who learns early that regular exercise is an important part of daily life, for example, might be considered to have acquired a valuable attribute, insofar as that knowledge and the behavior it produces make a significant contribution to the child's expected health and longevity. More broadly, society relies on citizens who take personal responsibility for their actions, respect the rights of others, and so on. Parents who take steps to impart these values to their children also might reasonably be considered to have added to their child's human capital.

Beyond these examples, there are skills that may contribute significantly to a person's enjoyment of life but might not be considered "productive" in the usual sense. For example, how should one think about the acquisition of the skills required to play bridge, windsurf, or appreciate a fine operatic performance? There are arguments on both sides as to whether the development of these skills should be considered an investment in human capital. In this report, we generally have drawn a distinction between the production of outputs for consumption and the enjoyment associated with consumption itself, with the latter not generally considered an output we seek to measure. This perspective might argue against treating the acquisition of recreational skills as an investment. The fact that there is a market for instruction in all of these areas—that people are willing to pay for the acquisition of these skills—might suggest the opposite conclusion.

THE HUMAN CAPITAL PRODUCTION FUNCTION

From birth to age 5, children develop a foundation on which their subsequent development builds. There is ample evidence that the home environment has strong effects on children's preparation for learning later in life. Families' preschool-age investments in what could be called childhood capital set the stage for the efficiency with which schools are able to educate students, for the amount of on-the-job learning people do after joining the work force, and for other kinds of learning that takes place later in life. Like the basic research and development that is a necessary precondition for having technology to embed in new machines, familial investments in young children—or some substitute for those investments when the family cannot or does not make them—are a precondition for having children capable of learning and eager to learn as they approach school (see National Research Council, 2000). The same argument applies to the complementary production of habits, tastes, and attitudes that affect people's willingness and capacity to produce good health, which in turn influences the effectiveness of the health care sector.

Many parents combine some paid care with caring for their children themselves. The effects of having young children in day care is a subject of ongoing debate. Based on research to date, there is an emerging consensus that long hours of nonmaternal child care for children under the age of 3 can have adverse effects on children's development, but much depends on the quality of that care relative to maternal care, which itself is highly variable across individuals (see National Institute of Child Health and Human Development Early Child Care Research Network, 1997; Love et al., 2003).

As any parent or schoolteacher can attest, there are important complementarities between the efforts of schools and the efforts of families in encouraging and assisting school-age children to learn. Quite distinct from what parents do before schooling begins, their ongoing encouragement also affects the ability and willingness of children to learn in school and the capacity of schools to teach children. Unlike the raw materials purchased by the manufacturing firm for its production of final goods, in the educational arena there is an ongoing relationship between the efforts of the firm—the school—and the supplier of that raw material—the family.

One indication of the widespread acceptance that environment matters is the support that programs like Head Start have enjoyed. The recognition that certain 6-year-olds may need remedial attention to become "school ready" implies that there are many 6-year-olds who do not need such remedial attention because they received the support needed to become school ready from their families. Spending on the Head Start Program is considered a part of national economic output. At least in concept, so too should be the familial efforts that prepare so many young children for their school experience without the need for compensatory preparation.

The discussion thus far might suggest that parents' influences on their children are the result of deliberate decisions. In fact, many of the effects that parents—and later, schools—have on children's preferences and habits are better described as spillovers than as explicit decisions (England and Folbre, 2000). Children growing up with highly educated, economically successful parents enjoy environmental advantages that are independent of the level of parental effort and attention they receive. Parents who want to set a good example for their children may be limited by their own shortcomings. Similarly, public schools can control their curriculum, but not the characteristics of their student body that affect motivation and readiness to learn. Many studies of "ethnic capital" and "social capital" describe environmental impacts over which neither parents nor children have much control (see Brooks-Gunn et al., 1993; Wilson, 1995). Small initial differences in capabilities can have feedback effects that lead to much larger differences, augmenting inequalities (Lundberg and Starz, 2000). These effects have implications for the measurement of relationships among the home environment, education, earnings, and well-being. Genetic differences, too, may be important in explaining observed variation in students' interest and ability to learn.

There is a vast literature on this topic, but still an enormous amount that is not known. The educational psychology literature, for example, has not produced scale measures that quantify a child's learning receptivity. There are agreed-upon measures of "IQ" and of achievements in certain subjects, though even they are subject to criticism. Less developed are measures of a child's social or emotional capacity for learning—the eagerness, the curiosity, the habit or capability to concentrate, and so on, that are critical to success in the educational system. Creation of better measures of both the cognitive and the noncognitive dimensions of children's development is an important area for further research.

Human beings are complex entities, and human development is a multidimensional process. Wise (2004) suggests using "sentinel capabilities"—measurable genetic, environmental, and health variables that affect an individual's capacity to transform potential into functioning—as key indicators of a broader pattern of development. Whether or not this particular approach proves to be a productive strategy, research in this area would be an important step toward modeling and analyzing families' contributions to preparing their children for schooling and later life.

Using presently available measures, the child development literature has explored the relationship between the family environment and child outcomes. But this literature does not tell how—through what mechanism—parents' interactions with their children affect those children's learning receptivity. Research on "parenting" offers interesting distinctions among styles and suggests that certain styles are more or less successful in producing certain attributes of the children. Still, that literature provides no basis for specifying a production function relating child outcomes to parental inputs, if indeed the process is so deterministic. Further research devoted to identifying the causal effects of different parenting behaviors on children's cognitive and noncognitive development is warranted.

One strategy that has yielded useful information on related questions has been to exploit variations in outcomes between twins or siblings who are separated at birth and raised in different home environments. Differences in outcomes between siblings who share all or a portion of their genetic make-up but have experienced different upbringings may support stronger inferences about the effects of environmental differences than differences in outcomes between individuals whose genetic make-up differs in unknown ways. One practical problem is the difficulty of identifying and matching up pairs of twins separated at birth. Much of the available knowledge comes from studies that include twin pairs separated up to 2 years after birth, which, by way of the common early environment, introduces some "contamination" to the results.[1]

A further area for investigation is the relationship between the cognitive, emotional, and social development of children at the time they enter school and later educational and labor-market performance. A literature exists that attempts to relate developmental attributes of children at the time they enter school to later educational attainment, but there remains much that is not known.

> **Conclusion 4.1:** For all of the reasons discussed above, we do not believe that the state of knowledge at the present time is such that efforts to develop a comprehensive human capital satellite account would be fruitful. There is too much that is not known about the relationship between inputs of household time and other resources, about the interaction between household inputs

[1]In part because of the limitations of twins studies, researchers also have used the natural experiment created by adoption to examine the relationships between family environment and outcomes. The growing literature in molecular genetics (e.g., Plomin, et al., 2002) provides additional evidence about the determinants of children's cognitive and noncognitive development.

and the inputs of the educational system, the health care system, and so on, and about the measurement of the relevant human capital outputs. Thus, the basic research foundation needs to be strengthened before a human capital account can be contemplated. Specifically, researchers need to continue working (1) to develop better measures of children's cognitive and, especially, noncognitive development; (2) to identify the causal effects of different aspects of parenting behavior on child development; and (3) to better understand the relationship between children's cognitive and noncognitive development at the time they enter school and these children's later educational and labor-market outcomes.

While the deficiencies in knowledge are real, in answering questions about schools' effectiveness in promoting skill capital, it matters very much what parents and other family caregivers are contributing to the development of preschool-aged children. In seeking "responsibility" within public schools—and looking at the financial expenditures, the curricula, and the quality of teaching staffs, and then using test scores or other output measures to assess the effectiveness of those "inputs"—it is important to remember that important non-school "inputs" from the family or the home affect the measured outputs. This nonmarket investment in children's human capital both precedes the influence of the public school during their first 5 years of life, and coexists with the school's influence during every year thereafter.

Put somewhat differently, the product of the formal educational sector is fundamentally a collaborative product requiring the cooperation of students, which may be reinforced or undermined by families. When the effectiveness of a school is assessed by looking at student achievement, treated as a function of school inputs without adequately accounting for the considerable efforts of the students or the compensatory and complementary efforts of the family, the results are apt to be misleading (Michael, 1999). Just as the schools often take too much credit for the remarkable achievements of their students, they are also too often accused of performing poorly when the difficulties lie in the home, not with the school.

Similar points can be made with regard to the production of health capital. At young ages, children's diet and exercise, exposures to harmful and beneficial environments, and so on, are determined by their parents; recent literature also has emphasized the importance of the uterine environment. At older ages, children continue to be influenced by their parents' health attitudes and habits. The maintenance of health capital is an interactive process, involving individuals' own behavior to promote their health (or to put it at risk) and their use of market-produced medical care to prevent or remediate health reductions. But too often, health is discussed and treated as though it were purely the product of the health sector, rather than a joint product of market-provided goods and services and home activities.

Perhaps one of the most undervalued, and surely one of the least well-measured, aspects of nonmarket-produced human capital involves the personal habits of citizenship—acceptance of personal responsibility, respect for the rights and interests of others, a willingness to take on social responsibilities, and so on. Values of ethical conduct, principled guidance from religious teachings or social commitment, empathy and civility, all are somehow and to varying degrees encouraged in youths. Through exposure to examples, admonitions and guidance from parents and caregivers, and observation and experience among their families, their ethnic, religious, and neighborhood communities, and their peers, children learn, they make judgments, and they form habits that influence behavior later on. These attributes have considerable influence on subsequent skill- and health-capital formation and on individuals' personal well-being and social contributions. In principle, the efforts and expenditures of resources by parents and families, and by nonmarket institutions like social clubs and religious or community organizations could be accounted for in some objective fashion. There is little dispute that these inputs and their outputs have value, but this is something we do not know how to measure.

Among the benefits of good child-rearing is that, if and when a child becomes a parent, he or she is likely to perform more effectively in that role. There is evidence, for example, that children's cognitive development, subsequent educational attainment, and health status all are positively related to their parents' education (Haveman and Wolfe, 1984, 1995; Wolfe and Haveman, 2001, 2002). Similarly, the home environment has an important effect on children's emotional and social development (National Research Council, 2000). In valuing the investment in a child's human capital, surely one would want to account for the enhancement of that child's capacity for raising productive, healthy children in his or her own turn, though again it is most unclear how one might do this.

A further point to be made (see Grossman, 1999) is that different types of human capital influence the production of other types within the nonmarket sector, particularly for education and health. It is probably the case that healthier people learn more easily and surely the case that more skilled or more educated people are more efficient in producing good health. The interrelationship between skill and health emphasizes yet again that the market sector production of these two products is not separable from the nonmarket production.

One could argue that simplifying assumptions are a necessary feature of any national income accounting framework. But one implication of this discussion is that many empirical estimates of the contribution of formal education to national income may be misspecified in two countervailing directions. In one direction, failure to account fully for the costs or benefits of parental inputs of time and money into the production of children's capabilities may overstate the rate of return to formal education. In the other direction, the persistent tendency to value education entirely in terms of its contribution to market earnings may understate its total contribution, since the increased education of parents is surely a major factor both in their own direct nonmarket production and in the subsequent development of their children's capabilities (Leibowitz, 1974).

FAMILY INPUTS TO THE DEVELOPMENT OF CHILDREN'S HUMAN CAPITAL

We concluded above that our knowledge base is not yet sufficient to warrant development of a comprehensive human capital satellite account. Having so concluded, however, we believe there is merit in considering what can be said about the magnitude of household investments in children, if for no other reason than to understand what implicitly is neglected when researchers focus on the production of human capital in the educational sector, the production of health capital in the health care sector, and so on.

Production of a child possessing any level of human capital typically requires both out-of-pocket expenditures on behalf of the child and inputs of parental time. As is true with other sorts of home production, out-of-pocket expenditures on children already are reflected in the national income and product accounts. What is not reflected is the time that parents and other family members devote to children.

Time-use studies from a number of countries indicate that married or cohabiting mothers of children under age 5 average between 2 and 3 hours a day in primary child care, depending on their hours of paid employment, while fathers average about 1 hour a day (Gauthier et al., 2001). In these studies, primary care is defined as time during which caring for a child is the primary activity in which a parent or other adult is engaged. If the reported figures seem small, it is because primary care represents only a small share of total time devoted to (or constrained by) children. Family members often care for children while simultaneously engaging in other activities, such as preparing meals or watching television. Survey results from Australia, a country that collects detailed information on secondary activities as part of its ongoing time-use data program, show that counting secondary activities more than doubles the hours devoted to child care (Ironmonger, 2003). The need to be available or "on call" also imposes significant constraints on parental schedules. Like firemen who spend relatively little of their time fighting fires, parents stand ready to provide attention, even when children are sleeping.

In recent cognitive pretesting for the American Time Use Survey (ATUS), the Bureau of Labor Statistics found that participants "strongly suggested that the concept of secondary child care is not intuitively meaningful, because most parents would consider those activities, 'just part of being a parent'" (Schwartz, 2002, p. 35). That task, "being a parent," extends well beyond tallies of activity hours. Statistics Canada opted for a national time-use survey that omits consideration of secondary activities but includes stylized questions regarding care time. The Bureau of Labor Statistics adopted a similar strategy for the ATUS, asking each respondent to estimate how many hours during the survey day children were "in their care" (Schwartz, 2002). Yet even this approach restricts respondents to time periods in which the adult and at least one child were awake. Similarly, the Child Development Supplement of the Panel Study of Income Dynamics (PSID), another source of data on time use, in this case from the child's perspective, collects information on time children spend with adults, but not directly engaged in activity with them. In this survey, too, children's sleep and personal care time are excluded from consideration.

Given the limitations of available data, most empirical studies of time devoted to parenting focus solely on primary activities. Still, the results are interesting and reveal some common patterns internationally. The University of Essex Multinational Time Use Survey data set includes data on time-budget surveys from more than 20 industrialized countries. These data show that increases in maternal employment cause only a small reduction in maternal time in activities with children. Moving from out of the labor force into a 35-hours-per-week job is associated with a decrement of only 7 hours a week in mothers' primary child care time (Gauthier et al., 2001). Studies of the more limited time-use data sets available for the United States confirm that maternal employment has surprisingly small effects. They also show that fathers' time with children remains low regardless of such factors as their own or their spouses' hours of work (Bianchi, 2000; Hofferth, 2000; Hofferth and Sandberg, 2001; Sandberg and Hofferth, 2001).

What has happened in the United States to parental time in activities with children as mothers have increased their employment? The data available to answer such a question are far from perfect (Budig and Folbre, 2004), but a small survey administered in 1998-1999 provides useful data for comparison with the results of an earlier 1965 survey. One recent study making use of these data (Bianchi, 2000) reports that mothers were engaged in primary activities with their children aged 18 and under for about 12 hours per week in 1998, compared with about 10 hours per week in 1965, a statistically significant increase. Time devoted to secondary activities with children also increased significantly, from just over 15 hours to just over 19 hours per week. The broadest measure, maternal time spent with children, increased only slightly, from about 37 hours per week to just over 38 hours (Bianchi, 2000, p. 405; Sayer et al., 2003).

Somewhat surprisingly, none of the measures show a significant decline in the time mothers devote to their children. One explanation for these findings might be that time with children is a sufficiently high priority that, as mothers' hours of paid work have grown, other activities have been sacrificed. Another, perhaps complementary, explanation might be that changes in household technology have made it easier than in the past to combine child care time with other activities. The availability of prepared foods that need only to be heated before consumption, for example, might make it easier to combine child care and meal preparation than would have been the case when meals had to be prepared from scratch; more widespread use of home computers might make it easier to combine child care with paying bills or carrying out job-related tasks; cable television and DVD players might make it easier to combine child care with entertainment activities; and so on. It is clear, however, that the choice of the appropriate definition of "child care" affects interpretation of trends.

A related issue is the quality or density of care time. In addition to parents sometimes combining other activities with supervision of their children, an hour of a parent's time can be devoted to the care of one or several children, and it may or may not overlap with the presence of other adult caregivers. Interestingly, because fertility rates have fallen over the course of the century, even if maternal time devoted to child care remains roughly constant, the amount of parental time spent per child may actually have increased (Bianchi, 2000). These issues bear directly on how the monetary value of parental time inputs should be estimated.

Analysis of PSID time-use data for 1997, based on time diaries for about 2,800 children under the age of 13, shows that children spent about 47 percent of their time sleeping or engaging in personal care. Of the time they spent participating in activity with another person—that is, of the 53 percent of their time they were not sleeping or engaged in personal care—about 45 percent was spent with a parent or a relative who probably was not paid. Much of this time involved overlaps: of the time they spent in activities with at least one parent, at least 32 percent involved the participation of another parent or family member and 22 percent involved the participation of a sibling. The PSID child development supplement diaries permit a clear matching of which child was getting parental input, but still have the problem of a parent spending time with other siblings at the same time. The average child aged 12 or under spent about 29 hours per week in activities with a parent (Folbre et al., 2004).

PSID data also show that parents invest in their children in a reactive fashion. Those who receive negative feedback on their child's performance or functioning have been shown to invest more in those children—one example being extensive investments of parental time to obtain some developmental threshold for a child with Down's syndrome. In the PSID child development supplement sample, even parents of children with less severe disabilities have been shown to devote more parental time to children reported to have physical limitations (Stafford, 1996). Who invests and why matters for the interpretation of data used to study parental time investment and early human capital development.

As already noted, the new ATUS will provide a wealth of information on how Americans spend their time, including information on the time that parents spend with their children. As noted in Chapter 2, however, the ATUS covers only the population aged 15 and older; it does not provide ongoing data on *children's* time use, of the sort collected as part of the PSID. Such data are likely to be important for a full understanding of how families affect children's development, and thought should be given to how they might best be collected.

VALUING THE TIME PARENTS DEVOTE TO CHILDREN

The "cost" of a child is the subject of a moderately large literature. The U.S. Department of Agriculture regularly provides estimates of spending on children based on analysis of the Consumer Expenditure Survey. These estimates are used to help define reimbursement rates for foster care, as well as the child support responsibilities of noncustodial parents. In 2000, parents in middle-income, two-parent families with two children spent an average of $8,915 a year per child under age 12 (Lino, 2001).

Significantly, these figures omit the value of parental time inputs, but available time-use data help to quantify that time. The PSID, for example, provides estimates of the average amount of time parents in two parent families with two children devoted to children under age 12 in 1997. Much depends on how child care time is defined. Using a relatively narrow measure that includes only activities in which both child and adult are involved, children received 31 hours of nonoverlapping parental care per week. By the broadest measure—counting all parental and other family time in activities with children plus supervisory time (defined as all time that children were not in the care of another person and not sleeping)—children received 79 hours of family care per week.

In Chapter 3 we argue that for most forms of household production the appropriate wage rate to use in valuing time inputs is the quality-adjusted replacement wage. If parents' productivity in child care equaled that of paid child care workers, it would make sense to use an estimate of child care workers' wages to value parental time input. In 2000, the average wage for a child care worker was $7.43 per hour. Multiplying 31 hours of weekly child care (the narrower measure derived from the PSID) by this low wage rate provides a lower-bound estimate of the value of parents' nonmarket time of $11,977 per year, per child (Folbre, 2004). This number suggests that the cost of a child in two-parent, two-child, middle-income families, including both cash expenditures and the value of time, amounts to $20,892, more than twice as much as cash expenditures alone. While foster parents and divorced custodial parents may devote less time, on average, to young children than married-couple parents, their time expenditures surely also are substantial.

The dollar cost estimate just cited hinges on parental time being no more (and no less) productive than that of hired child care providers. Of all the activities of household production, however, time devoted to the care of family members is probably the least completely substitutable with replacement market inputs. Many families benefit from the assistance of paid caregivers for part of the day, but a child who is raised entirely by paid caregivers may be less likely than one who also enjoys consistent parental attention to develop into a productive, healthy adult. If so, the appropriate replacement cost to use in valuing parental time inputs into the care of their children, at least at the margin, could be significantly higher than the wage paid to child care workers. Given the present state of knowledge about the production function for human capital, broadly defined, however, it would be difficult to say what the right wage rate might be.

Another candidate wage rate for use in valuing parental time inputs would be the opportunity cost of that time, perhaps proxied by the parent's market wage rate. This is an alternative that we have rejected in the case of other sorts of home production, reflecting in part our belief that there is likely to be only a weak relationship between productivity in most home production tasks and market wage rates. In the case of parents' interactions with their children, however, there is evidence that parents with more human capital are more productive not only in the market, but also in fostering the development of their children's human capital. What we do not know much about, however, is the strength of the relationship between the productivity of parents' market time and the productivity of the time they devote to child-rearing activities.

As an aside, it may also be worth noting that the opportunity costs of parental time devoted to child care—or more broadly, taking responsibility for their children—may go beyond any potential earnings forgone as a consequence of spending a marginal hour in child care rather than market work. The commitment to bear and raise a child often has life-style implications—e.g., deciding to stop smoking or cease alcohol consumption during pre-pregnancy and pregnancy intervals, changing or postponing career commitments, or choice of location and attributes of the family home. It is difficult to know, however, how one would incorporate this insight into a family care satellite account.

One concern with using the opportunity cost of parents' time to value their contributions to their children's human capital is the difficulty, if not the impossibility, of distinguishing between the consumption and the investment components of this care; and we do not suggest that time-use surveys or satellite accounts should attempt to do so. Nonetheless, the relationship between family care and human capital has important implications for the interpretation of nonmarket satellite accounts for other sectors, most notably education and health.

5

Education

This chapter explores the possibility of constructing a satellite account for education within a framework that is compatible with the national accounts. During their younger years, individuals spend a large percentage of their nonmarket time engaged in school-related activities. The focus then tends to shift from formal to informal education and from school-based learning to work-related training as individuals become adults—but education continues. Because its benefits are realized over an extended period, education should be thought of as an investment in human capital. Early economists such as Adam Smith and John Stuart Mill recognized this fact.[1] Although the nation's gross domestic product (GDP) includes expenditures for education, it fails to capture fully the contribution of nonmarket time spent in education to future economic growth, the well-being of individuals, and society in general. Individuals with a higher level of education tend to earn higher incomes; have higher productivity in the workplace and elsewhere; be better informed about and more involved in community activities; and generally report that they are happier.

Satellite accounts for education would be useful to researchers and policy makers in at least three ways. First, nonmarket inputs such as time—including the

[1]In *The Wealth of Nations*, Smith (1776, p. 141) wrote: "The work which he learns to perform, it must be expected, over and above the usual wages of common labour, will replace to him the whole expence of his education, with at least the ordinary profits of an equally valuable capital. It must do this too in a reasonable time, regard being had to the very uncertain duration of human life, in the same manner as the more certain duration of the machine." See also Mill (1848, as quoted in Oser and Brue, 1988, p. 138).

time of students, parents, and others—and social capital are important aspects of investment in education, and they are missed in the traditional National Income and Product Accounts (NIPAs). Looking at all education inputs, and their related outputs, would form a more complete picture. Second, human capital, particularly that arising from education, is large relative to the nonhuman capital stock measured by the Bureau of Economic Analysis (BEA). By one estimate, more than two-thirds of national income in recent years is a return to past investments in schooling and to work experience (Krueger, 1999). Separate education accounts would contain data essential for improving our understanding of how investment and the capital stock, defined more broadly to include both human and nonhuman capital, affect economic growth. Third, the education sector is large and important in its own right. Understanding trends in output and productivity in the education sector, both public and private, therefore is of interest.

CONCEPTUAL FRAMEWORK

In this chapter we discuss the components of an education satellite account, including inputs and outputs, focusing primarily on formal education and the significant measurement issues it involves. There are difficulties on both the input and the output side: How does one value time that is not transacted in an explicit market? How does one define and value the output of education, given that it is not directly traded and not directly observed? Regardless of the methodology employed, a goal is to have independent estimates of inputs and outputs—both market and nonmarket—in nominal and constant dollars, to allow for the estimation of productivity.

The emphasis of this chapter is on formal education, in part because data on informal education and training are limited.[2] This is not to imply that the role of families in preparing young children for school early in life or on-the-job-training and other sources of informal education later in life are unimportant. Rather, we do not discuss these less formal investments only because their magnitude and effects are difficult or impossible to measure at present. The scope of the chapter also excludes other factors, such as health and the role of social capital, that may enhance or impinge on education.[3] By looking at formal education, the chapter highlights approaches that could be used to produce more comprehensive education accounts and measurement issues that would arise in the construction of such accounts.

The National Center for Education Statistics, the Bureau of Labor Statistics (BLS), and the Census Bureau, as well as the National Education Association and

[2]Schools and their staffs also provide child care services, which are valued by parents, but these services can be considered a separate joint product of having children attend school, rather than an output of education itself.

[3]Social capital is defined in Chapter 1; see Chapter 6 for a discussion of accounting for health.

other private organizations and researchers, have compiled a wealth of data on education. These sources are used by the BEA in constructing its education-related estimates. The BEA estimates expenditures for both government-provided and privately provided education. In general, only expenditures for market inputs are included, although an imputation is made for capital services from fixed assets (e.g., equipment, structures, and computer software).[4] Expenditures for meals, rooms, and entertainment are excluded. Quantities at the category level are generally estimated as nominal expenditures deflated by a weighted input cost index. Aggregates are then created with cost weights for government-provided education and revenue weights for privately provided education.[5] The prices of inputs to education generally are not adjusted for changes in quality, but there is an adjustment for changes in teaching staff composition.[6] The BEA index of public school teaching staff composition is based on an index of the number of teachers weighted with salaries cross-classified by years of experience and highest degree obtained. As the existing accounts do not measure output independently of inputs, it is impossible to estimate productivity in either the public or the private education sector. The BEA is now studying the feasibility of constructing output measures independently from input measures for public education (Fraumeni et al., 2004).

Information is available on market inputs to education, such as teachers, buildings, and books, but not on nonmarket inputs, such as social capital and time, including the time of students, parents, and others. Outputs cannot be measured easily. Counting numbers of students enrolled is not the same as measuring the amount of education received, although years of schooling is a strong predictor of earnings and other economic outcomes.

The value of nonmarket inputs to education could well be larger than market and government expenditures for education, estimated to be approximately $628 billion in 2001, though without specific estimates for the value of nonmarket inputs it is difficult to know for sure.[7] To begin conceptualizing how a satellite account might be structured to give a more comprehensive picture of education, it is useful to identify the relevant inputs and outputs, as is done in Table 5-1.

Market inputs include paid labor, materials, and fixed capital; nonmarket inputs include volunteer labor, students' and parents' time, and social capital.

---

[4]The capital services from these assets are set equal to consumption of fixed capital (depreciation). These imputed capital services are allocated to consumption.

[5]Estimates for education are included in the NIPAs and in the BEA's GDP by industry accounts. Some real estimates depend on indexes from sources other than the BEA, but these are input based measures.

[6]Quality adjustments are made for some inputs that are not specific to education, the most notable being the quality adjustments made to the price index for computers.

[7]See the discussion of the value of time that primary and secondary students devote to their education in Chapter 1. All BEA data in this chapter are of a pre-December 2003 comprehensive revision vintage.

TABLE 5-1 Education Inputs and Outputs, Market and Nonmarket

Inputs	Outputs
Paid labor	Educated individuals
Teachers and support staff	Higher income from being more educated, higher workplace and nonmarket productivity
Volunteer labor	Intangibles
Parent Time	Better informed citizens, improved individual and societal well-being
Student Time	
Materials	
Books and other	
Fixed capital	
School buildings and other structures, equipment, and computer software	
Social capital	

Students learn more with teachers, books, facilities, and help from others than without these inputs. Volunteer labor is listed separately as an input to capture the unpaid time spent by people other than students and their parents (see the discussion of volunteer labor in Chapter 7). Students bring to the learning process aptitude and attitude, the latter of which can be strongly influenced by the social environment in which they live. Family, community, and peer attitudes all clearly can influence a student's attitude toward learning.

The output of education is educated individuals who are more productive and thus earn higher incomes and also contribute to society in other, less tangible, ways. A possible measurement methodology is to estimate the value of education by the amount that an increase in education raises an individual's earnings. Higher productivity is also an output of education. If the time of more educated individuals is treated as a more valuable input than the time of less educated individuals, estimates of productivity will not rise as the workforce becomes more educated. But if labor inputs are not quality adjusted in this way, one of the effects of an increase in education levels would be an increase in estimated productivity in the sectors that employ more educated workers. It would be incorrect, however, to count both this increase in productivity and the higher earnings of more educated workers in determining the return to education. Intangible outputs, especially those that may accrue to society as a whole rather than to the individual receiving an education, are apt to be particularly difficult to measure.

An education satellite account can incorporate nonmarket inputs and outputs. In addition, there may be more flexibility to experiment with alternative measures of market inputs and outputs in a satellite account. Education occurs in a complex environment in which many elements may influence both the quantity and the quality of education received, and the benefits of that education for the recipients and for society. Accordingly, both conceptual and measurement issues must be confronted.

> **Recommendation 5.1:** An education satellite account, presented in nominal and real dollars, should be produced by the Bureau of Economic Analysis in collaboration with the National Center for Education Statistics (NCES).

MEASURING AND VALUING INPUTS

Market and government inputs to education can be estimated relatively easily, but estimating nonmarket inputs (other than those originating from the government) is difficult. A variety of data on market and government inputs to education are available, but data on nonmarket inputs are harder to find and much less complete.

Market and Government Inputs

Market and government inputs to education account for a substantial share of GDP: Table 5-2 gives a snapshot of the sources and uses of expenditures on education in 2001, and Table 5-3 gives a historical perspective. From the mid-1960s onward, education expenditures as a percentage of GDP hovered in the 5-6 percent range; this percentage was lower in earlier years. Spending for elementary and secondary education dominates other expenditures, accounting for approximately 70 percent of spending included in GDP. Not surprisingly, state and local spending is primarily for elementary and secondary education although, since 1970, about 20 percent of state and local expenditures are for higher education. Spending by state and local governments includes federal grants-in-aid to

TABLE 5-2 The Nation's Education Dollar, 2001 (percent of expenditures)

Sources		Uses	
Federal government	1	Primary and secondary education	71
State and local government	86	Higher education	22
Private education services	13	Other	7

NOTES: State and local government expenditures include grants-in-aid from the federal government that account for 4 percent of education spending. Shares were constructed with data from the National Income and Product Accounts (Tables 3.15 and 3.17) and GDP by industry data, available on the BEA website (www.bea.gov). Intermediate inputs to the private educational services industry, such as books and supplies, are excluded from GDP, while government expenditures for intermediate inputs are included in GDP.

TABLE 5-3 Education Expenditures, 1952-2000

			Expenditures as a Percentage of Total Education Expenditures						
			Sources				Uses		
Year	GDP (billions)	Total education expenditures as a percentage of GDP[a]	Federal government	State and local government	Within state and local, federal grants-in-aid	Private education services	Elementary and secondary education	Higher education	Other
1952	358.6	3	1	88	3	11			
1960	527.4	4	1	88	2	10			
1970	1039.7	6	1	87	7	12			
1980	2795.6	6	1	89	6	11	70	24	6
1990	5803.2	5	1	87	4	12	70	24	6
2000	9824.6	6	1	86	4	13	71	22	7

[a]Includes only expenditures reflected in GDP.

state and local governments. These grants-in-aid accounted for no more than 3 percent of education spending by state and local governments from the 1950s through the mid-1960s. This percentage at least doubled through the early 1970s, then dropped to at most 5 percent during the 1990s. Spending on higher education in the last quarter of the twentieth century accounted for 50-60 percent of the total private educational services spending included in GDP; the share of vocational schools and other "not elsewhere classified" spending increased to about 25 percent, and the share of elementary and secondary schools decreased to 20 percent over the last few years.

None of the BEA's GDP-by-industry estimates include all of the capital input costs associated with the provision of education. Information on a portion of the relevant investment and capital stocks, which could be used to estimate full capital costs, is available from BEA. It includes investment and capital stocks of the private educational services industry and educational structures held either by the federal or by state and local governments. The full capital service flow corresponding to these stocks, and others, could be estimated by making assumptions regarding the net return to capital; the depreciation component of the capital service flow is already included in GDP (see National Research Council, 1998). We do not detail the methodology that could be used, except to note that capital input to education is probably understated in GDP.[8] To allow for comparisons across time and for productivity estimation, a measure of the real inputs to education is needed. Such a measure could be readily constructed for market inputs using data from the BEA that are available as of the fall of 2004.

Since public education is an obvious near-market activity, it makes sense to measure inputs and outputs in the same way for public and private schooling. It would be useful to have consistent methodology and coverage for the government-provided and privately provided education categories in the national accounts. In addition, it would be helpful if estimates of GDP for private educational services were available by subcategory, disaggregated at least for primary, secondary, higher, and other education. Such disaggregation would allow for analysis of public and private education by level (primary, secondary, and higher education), as well as for all levels combined.

Another way to summarize direct monetary inputs to education is by type of expenditure—that is, salaries, capital costs, operation, and other. As already noted, public elementary and secondary schools make up the largest part of spending on education; Garrison and Krueger (2004) provide estimates for this category, broken down by expenditures on instruction, administration, plant maintenance and operation, fixed charges and other school services, and capital outlays. The largest expenditure is for salaries—the main component of the instruction and administration categories.

[8]Murphy (1982) estimates the capital service flow (cost of capital services) for school structures owned by governments and nonprofits institutions; he assumes that the net return is positive.

Nonmarket Time Inputs

In this section we focus on educational activities directly related to formal schooling. These include students' time spent in school and doing homework, as well as parents' time spent in school-related activities, such as parent-teacher meetings and assisting their children with homework. Other types of educational activities, such as attending plays, reading books for leisure, or on-the-job training, are not considered.

Time Estimates

The most promising sources of information on time spent in educational activities are time-use surveys, as only limited information is available from other sources. Eisner (1989) used the Michigan surveys of 1965, 1975, and 1981 for that purpose (see, for example, Juster, 1978), but does not report the actual time estimates. Gates and Murphy (1982) and Jorgenson and Fraumeni (1989) both relied on the Coleman report (Coleman et al., 1966) for estimates of time spent in school and homework. Jorgenson and Fraumeni assumed that all enrolled students spent 1,300 hours per year on these activities over the 1948-1986 period they studied. Gates and Murphy (1982) do not report their assumption about hours per student. Garrison and Krueger (2004) estimate the time spent in public elementary or secondary school as 6.5 hours per day times the term length. They vary the term length from 175 days in 1939-1940 to 180 days in 1999-2000. It should be noted that, unfortunately, BLS's new American Time Use Survey does not provide data on the allocation of time for children under the age of fifteen, and thus cannot be used to estimate the time they devote to education.

Eisner (1989, p. 66) estimates average minutes per week by adults in child education at home, by sex and employment categories. Data are available from the Michigan time-use surveys for selected years; these data are extrapolated to cover the whole period of interest. Minutes per week vary from a low of 5 minutes for nonemployed males in 1981 to a high of 116 minutes for nonemployed females in 1946-1975. Neither Gates (1984) nor Jorgenson and Fraumeni (1989) separate out time spent by parents in school-related activities, and Garrison and Krueger (2004) do not include parent time devoted to child education in their estimates. More and more up-to-date information on student and parent time inputs to education clearly is needed.

> **Recommendation 5.2:** Information on time-use patterns from statistical agencies and other sources, together with a consistent set of demographic data, should be used to estimate the amount of time spent in school and in educational-related activities by students and their parents. New sources of information will be needed for these estimates.

Time Valuations

Several different approaches have been developed to value the time spent in educational activities; they generally use an opportunity cost concept to estimate the value of students' time inputs to education and a replacement cost concept to estimate the value of parents' time inputs to education. The size of the estimates for the value of time inputs varies significantly as currently there is no consensus on best methodologies.

Kendrick (1974), Eisner (1989), Garrison and Krueger (2004), and Gates (1984) estimate students' time cost as the compensation pupils could have earned in a market job. Kendrick assumes that only students aged 14 and older could perform market work; accordingly, the opportunity cost for younger students is zero. Eisner uses an updated version of the Kendrick data. Gates assumes the opportunity cost for students to be positive only for those aged 16 and older. All of these studies use estimates of actual wages paid to measure opportunity cost. Garrison and Krueger demarcate the positive and zero opportunity cost groups by grade level. For students attending grades 9 through 12, the opportunity cost is assumed to equal the minimum wage; for students in lower public school grades it is assumed to be zero. One relevant consideration in assigning these valuations is that child labor laws prevent young children from working. Accordingly, it seems reasonable to assume as an approximation that young children face an opportunity cost of zero for the time they spend in school, even though children younger than 14 could perform some useful work and did so to a much greater extent in earlier time periods (e.g., farm work and household tasks).

Jorgenson and Fraumeni (1992) argue that a positive value should be placed on the time of younger students because enrollment in higher grades is predicated on completion of lower grades, but they value this time based on the future earnings stream it produces rather than the current earnings foregone.[9] More specifically, Jorgenson and Fraumeni calculate the value of students' time input to education as a residual—the difference between the estimated value of investment in education and the value of market inputs. Because the value of investment is so large and the value of market inputs is relatively small, the residually determined value of students' time input to education is very large.

Opportunity cost approaches have also been explored for valuing parents' time. Gates (1984) and Murphy (1978) estimate such values for parents' time inputs to child care and instruction. Jorgenson and Fraumeni (1989) use an opportunity cost approach to value parents' time inputs to education. Alternatively, parent's time inputs to education could be treated as an investment by parents in their children, using an approach similar to the one they took for valuing investment in education by the students themselves.

---

[9]Other researchers who use a present value approach include Walsh (1935) and Graham and Webb (1979). The research cited by Jorgenson and Fraumeni was begun by Jorgenson and Pachon (1983a, 1983b).

For valuing parental time inputs to education (e.g., the time parents spend helping students with homework), valuations based on replacement options are more common than those based on opportunity cost. Some researchers (e.g., Eisner, 1989; Landefeld and McCulla, 2000) have used a generalist wage, while others have used a specialist wage. Murphy (1982) uses the average wage paid to domestic workers to value hours spent in child care and instruction, but also offers alternative estimates of the value of parents' time inputs using a specialist wage. His specialist wage is a weighted average over five occupational categories: child care provider in private household, child care provider outside household, welfare service aide, school monitor, and teacher aide. Presumably, only the last two categories would be included in the weighted wage average for valuing parents' time input to instruction on its own. Although Murphy's parent time estimates are available only for 1976, they are of particular interest because he also presents estimates for three variants of the opportunity cost approach (gross compensation, after-tax compensation, and net compensation) to value time inputs. The generalist and specialist approaches both produce lower estimates for the value of parents' time than the opportunity cost approach, reflecting the lower-than-average wage rate paid to housekeepers, child care providers, and teacher aides. Garrison and Krueger (2004) offer an alternative perspective: they estimate the amount that parents of public school children would pay to have their children cared for during school hours if the children were not enrolled in school, but not the time cost of providing instruction to their children at home. They make use of cost figures for child care from the Census Bureau.

One criticism of all three approaches to valuing parents' time—the generalist (sometimes labeled housekeeper), specialist, and opportunity cost approaches—is that, in practice, average market wages, not marginal wages, are used to construct estimates. Using a replacement cost approach and valuing the parents' time at the specialist wage is sensible if one believes that parents' expertise is roughly equivalent to that of the average specialist. A preferable option, however, would be to use a productivity-equivalent replacement wage that accurately reflected the relative skill of a parent relative to that of the specialist in providing educational services. An opportunity cost measure may overvalue the parent's time spent in providing educational services, particularly for a high-wage parent. The difference between a specialist (or generalist) wage and the productivity-equivalent replacement wage is probably less on average than the difference between the opportunity cost of parent time and the productivity-equivalent replacement wage.

There are other questions about whether a market wage adequately reflects the value of the services being performed. First, the wages of the relatively few specialists who are paid to perform activities primarily carried on outside of the marketplace arguably would be higher if all of these activities were market activities, as the demand for them would be higher (Murphy, 1978). Second, the market wage may not reflect the full value of these activities because of nonpecuniary returns enjoyed by those who perform them or other factors. It is difficult to know how one should adjust the market wage to adjust for these last two considerations. The panel believes that productivity-equivalent replacement cost provides the best conceptual basis for valuing parent time spent in child education; in practice, it would be extremely difficult to adjust candidate replacement wages for productivity differences between parents and market providers.

Most researchers calculate the total cost of nonmarket time devoted to education—students' and parents'—as the sum of two positive costs. In Garrison and Krueger (2004), however, parents' time costs can be negative because they include the potential savings of child care for parents while students are in school. Specifically, they count the amount parents save on child care when a child is in kindergarten through grade 6. The time they (or someone else) otherwise would have had to spend caring for the child is valued at the average cost of child care, using a specialist approach. In this sense, it makes no difference whether the child would have been cared for by a parent or by someone else, such as a child care professional. The cost savings for young students offset the non-negative opportunity costs incurred by older students enrolled in school; accordingly, the nonmarket time costs of education may be negative, depending on the age distribution of the students by grade.

As shown in Table 5-4, it is clear that the magnitudes of the estimates of time costs can be significantly different, even when the same general approach is used. Both Gates and Eisner use an opportunity cost to value students' time, but the Eisner estimate for 1979 is more than 50 percent larger than the Gates estimate. This difference is due in part to the range of years for which opportunity cost is estimated: Gates calculates the cost for students aged 16 through 24, while Eisner calculates it for students aged 14 through over 35. The largest differences shown are those between the Jorgenson and Fraumeni estimate and all of the others. As described more fully below, Jorgenson and Fraumeni impute the value of student time as a function of the incremental addition to future lifetime income from completing another year of school. Garrison and Krueger's estimates of the value of student plus parent time inputs to education are the smallest. This result certainly reflects the coverage of their study (K-12), but it also reflects their counting parents' time not spent on child care for young children during the school hours as an offset to the cost of the investment. As would be expected, adding college education to the Garrison and Krueger calculations significantly increases their estimate of input costs (see the last column of the table). Regardless of the specific methodology used, it is important to recognize that education normally requires commitments of time from both students and parents.

Recommendation 5.3: The value of the time that students and parents spend in educational activities should be estimated and included in satellite accounts. In principle, the time cost for students should be estimated using a modified opportunity cost approach. Parents' time should be valued as the productivity-equivalent replacement cost of providing comparable educational services.

TABLE 5-4 Value of Time Inputs to Education in Selected Studies (billions of nominal dollars)

Year[a]	Gates[b] (Students)	Eisner (Students)	Eisner (Students and Parents)	Jorgenson-Fraumeni (Students)	Garrison-Krueger (K-12 Students)	Garrison-Krueger (K-12 Students and Parents)	Garrison-Krueger (K-16 Students and Parents)[c]
1940					2.083	–1.176	
1950		16.420	20.577	255.353	3.703	–2.243	
1960		33.320	40.721	644.510	9.872	.191	7.664
1970		103.857	117.933	1365.691	24.673	9.218	36.039
1980	161.5[c]	259.918	285.911	2289.584	45.156	19.684	92.240
1990				3686.547	45.525	–1.906	125.227
2000					79.332	8.808	203.112

[a]Calendar year, except for Garrison-Krueger entries, for which it is the school year that ends in the indicated year.

[b]For the year 1979.

[c]Includes public elementary and secondary school students and their parents, and all college students. The opportunity cost of college students' time is valued at the average wage of production, nonsupervisory workers on nonfarm payrolls.

NOTE: See text for citations.

Schools and their staffs can be viewed as producing two outputs: education and child care services. In the framework recommended here, the value of investment in education would not include the value of child care services provided to young children while they are in school. Both education and child care services are investments in human capital. For the account being considered in this chapter, however, we are concerned with activities directly linked to the production of educated individuals. Accordingly, while a satellite account for human capital (broadly defined) would estimate the value of the child care services schools provide as well as the value of the education they offer, an education satellite account would include only the latter as output.

MEASURING AND VALUING OUTPUT

Researchers do not typically attempt to produce direct measures of the output of the education sector. Frequently, the value of educational output is set equal to the cost of educational inputs, and no price is estimated for output or inputs. Accordingly, little is known about growth, quality improvements, or productivity in the education sector.

Educational output has been or could be directly estimated using any of three general approaches—the indicator approach, the incremental earnings approach, or the housing value approach—each of which we discuss in turn. Regardless of the approach used, it should be recognized that a significant component of education is an investment in human capital. In the NIPAs, education expenditures other than those for structures and equipment are treated as consumption or intermediate inputs, rather than as investment. The output of the education sector is the flow of improved capabilities that result from schooling each year. That accumulation of skills adds to the stock of human capital.

Recommendation 5.4: Education should be recognized as an investment in human capital in a nonmarket satellite account.

The human capital created through education can have both private and social benefits. Human capital is a broad term meant to capture the skills, attitudes, and abilities that are valuable because they raise productivity in market and/or nonmarket activities. Human capital can lead both to private benefits that accrue only to those who acquire the education and to external benefits that accrue to others. In addition to yielding higher individual earnings, for example, education might increase voter participation, lead to voters being more informed, and result in reduced crime.[10] Any of these positive externalities is a potential source of social benefits from education. Nelson and Phelps (1966) and Romer (1990) are

[10]See Haveman and Wolfe (1984, 1995) and Wolfe and Haveman (2001, 2002) for a catalog of the non-earnings outputs of education.

among the many researchers who have argued that the social return to investment in education exceeds the private earnings return. Others argue that the private return to education exceeds the social return because of signaling. In this view, education is seen as a costly mechanism that serves to sort workers in terms of productivity without necessarily enhancing their skills.[11]

The available empirical evidence points strongly to sizable private benefits from education. Extensive research documents that individuals with more education tend to have higher earnings (see Card, 1999, for a survey). Each additional year of schooling is associated with roughly a 10 percent increase in earnings. It is possible that those with a high level of education tend to have more inherent ability or higher motivation, which would have led to their having higher earnings, on average, even without their high level of education. If this were the case, the observed education-earnings gradient would be an upwardly biased estimate of the causal impact of education on earnings. Much of the available evidence suggests, however, that the higher earnings associated with higher education result from increased cognitive and noncognitive skills due to the education *per se*, not from other unmeasured factors that are correlated with education.

A number of studies address the possible effect of selection on estimated returns to education by exploiting exogenous variation in educational attainment due to differences in compulsory schooling requirements or geographic proximity to a college (see Card, 1999). Other studies look at the differences in earnings between identical twins with different levels of education to net out unobserved family and individual factors. Ashenfelter et al. (1999) compiled estimates from 27 studies representing 9 different countries, and found that, if anything, omitted variables tend to lead researchers to understate the private return to education.

The evidence on the magnitude of externalities from education—either positive or negative—is less clear cut. In practice, the importance of externalities has been very difficult to assess. Many implications of the sorting model are similar to those of the human capital model, so it has proved difficult to distinguish between the two perspectives, although studies of the effects of changes in compulsory schooling requirements provide some support for the human capital model. In principle, the strongest evidence regarding the existence of educational externalities could come from studies that estimate how differences in educational attainment across countries relate to GDP or how increases in education over long periods of time relate to GDP growth. If education has a larger (smaller) effect on GDP growth than would be expected based on estimates of the private return to education, this could be taken as evidence of positive (negative) externalities from education. Cross-country evidence is always difficult to interpret because there are only a relatively small number of countries and many potential

[11]Spence (1974) develops a model in which education signals that a person is of higher ability; Alison Wolf (2002) argues that this type of credentialing is important.

influences on GDP and because there are difficulties with comparing educational attainment across countries. Nonetheless, we interpret the bulk of the evidence as indicating that increases in education are associated with higher living standards because education raises individuals' productive capacities (see Krueger and Lindahl, 2001; Cohen and Soto, 2002; Heckman and Klenow, 1997; Acemoglu and Angrist, 2001; for a different view see Benahabib and Spiegel, 1994).

Indicator Approach

The simplest indicator approach to measuring the output of education makes no adjustment for changes in the quality of education or for changes in the productivity of the educational sector over time.[12] Often it takes the total number of students or (better) the total number of hours spent by all students in educational activities as the output index. A problem with the simplest indicator approach is illustrated by considering the impact on output estimates of a reduction of a given percentage in class size with a corresponding decline in the total number of pupils in the system and no change in the teaching staff or facilities used. The output of the education sector, and possibly productivity, will be estimated to have decreased by the same percentage, even though the quality of education provided (as measured by the skill level achieved) probably will have increased.

More sophisticated indicator approaches use student test scores, subjective ratings of the overall quality of instruction, degrees granted, the number of students advanced from one grade to the next, or other indicators. Hoxby (2003), Garrison and Krueger (2004), and others use test scores as a measure of performance for the U.S. education system. Test scores have not been adopted into the U.S. national accounts,[13] though some researchers working on foreign national accounts have used test scores and other indicators.[14]

In the remainder of this section, we outline an example of the output indicator approach that incorporates performance measures. This example could be extended to more accurately assess the "value" of output as proxied by test scores by looking at the relationship between scores and earnings or scores and housing values; at this point, however, this next step has not yet been taken.

---

[12]Inputs can increase if more of the same type are used (e.g., if more teachers are hired), or if the quality of inputs increases (e.g., by teachers receiving more training). Output can increase if more students are being educated or if the quality of output increases (e.g., because they are learning more). Productivity is measured by output per unit of input; accordingly, productivity estimates require information on output and inputs. If an output is estimated by an input index, zero productivity change is being assumed.

[13]Researchers at BEA are looking at measures of education output for incorporation in the next NIPA comprehensive revision, which is scheduled for 2007 or 2008.

[14]Fraumeni et al. (2004) summarize indicator approaches used by the statistical agencies in a number of countries.

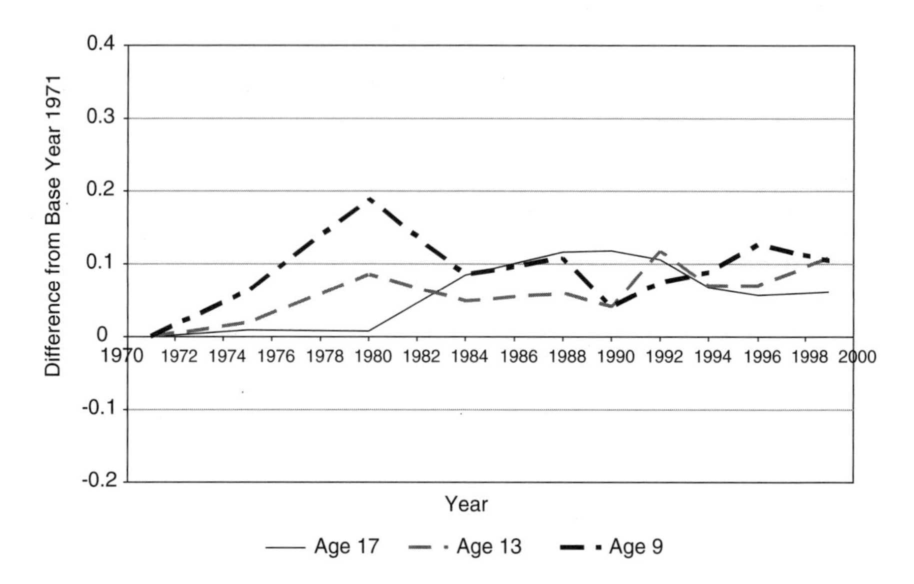

FIGURE 5-1 NAEP reading scores, 1971-1999.
NOTE: Difference is calculated as a percentage of cross-sectional standard deviation (1996).
SOURCES: NAEP 1999 Report, Trends in Academic Progress, Table B1. NAEP 1996 Trends in Academic Progress, Tables C16-C18.

The most commonly used measure of school performance in the United States is the National Assessment of Educational Progress (NAEP). NAEP is a nationwide test administered to 9-, 13-, and 17-year-olds that provides time-series data on reading, mathematics, and science achievement. Hoxby (2003), for example, uses NAEP points per dollar spent as a measure of school productivity. Whether NAEP scores map linearly into output is an open question.

Garrison and Krueger (2004) look at trends in NAEP subject scores to study changes in the quality of education output. Figures 5-1 and 5-2 show trends in average NAEP scores for two of the three subject areas (reading and mathematics).[15] Scores have been normalized into units of standard deviation changes from the initial year in which data are available. The standard deviation is from a single cross-section.

[15]A figure for the NAEP science scores is not shown because of difficulties with comparing the scores over time.

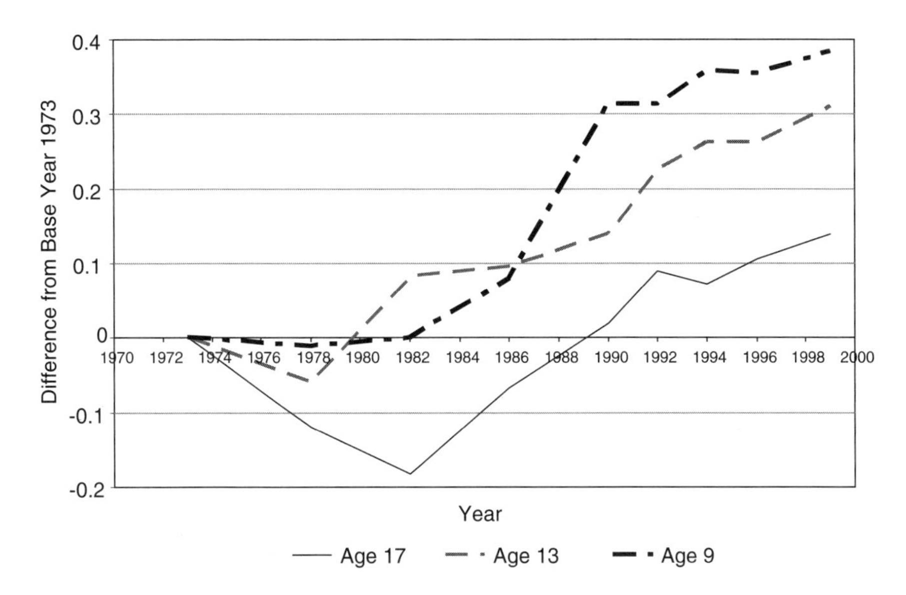

FIGURE 5-2 NAEP mathematics scores, 1973-1999.
NOTE: Difference is calculated as a percentage of cross-sectional standard deviation (1996).
SOURCES: NAEP 1999 Report, Trends in Academic Progress, Table B1. NAEP 1996 Report, Trends in Academic Progress, Tables B16-B18.

Using NAEP scores as a measure of performance has several limitations. First, NAEP data are available for only a relatively short period of time. In addition, NAEP scores do not account for post-high-school gains (e.g., improvements in cognitive skills due to higher education, on-the-job training, etc.) and do not assess skill retention. Therefore, NAEP scores do not reflect the *current* cognitive skill of tested cohorts. And, the separate NAEP scores would have to be assigned weights to produce a single output indicator series. Test score growth rates then could be calculated and applied to the nominal dollar input expenditure in the base year to form a quality-adjusted output indicator suitable for a dollar denominated satellite account.[16]

---

[16]An output index can adjust for change in output quality and change in the number of students. If output quality is normalized to 1.0 in the base year and multiplied by the number of students, an index is created which adjusts for both types of possible changes. Growth rates can then be calculated from this index and applied to the nominal dollar input expenditure in the base year to form a quality-adjusted output indicator.

Data from the International Adult Literacy Survey (IALS) could be used to supplement the NAEP scores. Designed for cross-country comparisons, this household-based survey of adults (aged 16-65) was performed in 22 countries between 1994 and 1998. The survey collected demographic data and work history information and assessed literacy by administering a standardized test. For each respondent, the IALS provides three estimates of literacy—prose, document, and quantitative—scored on a scale of 0 to 500. The IALS data can be used to create cohort-level trends in test scores.

The IALS scores reflect the current cognitive skill of adults in the participating countries. Since all three literacy measures are very highly correlated (see Blau and Kahn, 2001), it is reasonable to focus on a single measure. Figure 5-3 shows the mean IALS prose score by year of birth (5-year moving average) for the United States (natives only), United Kingdom, and Sweden.

Garrison and Krueger use IALS scores by birth cohort to make inferences about how the output of the education system has changed over time. Drawing inferences about cohort trends from a single cross-section is complicated because aging can affect the cohort scores independently. Additionally, IALS results provide a broader measure of human capital than that associated with formal education alone. To construct meaningful time-series flow data from these results,

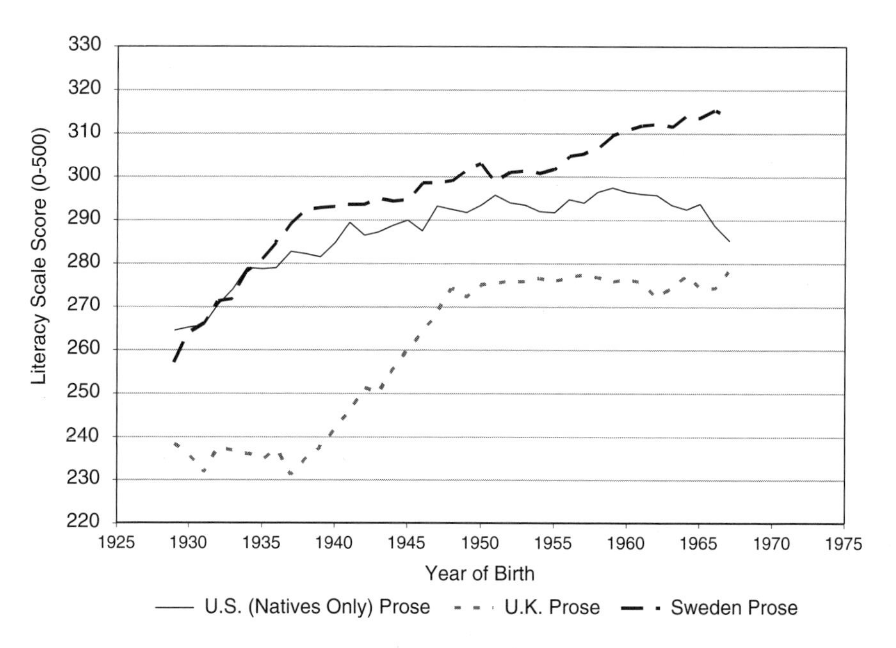

FIGURE 5-3 IALS prose scores by year of birth for the United States, United Kingdom, and Sweden (5-year moving average).
SOURCE: Organisation for Economic Co-operation and Development, 2000.

the effects of aging and other human capital enhancing factors would need to be accounted for. Nevertheless, repeated cross-sections of the IALS could be used to compute measures of the stock of human capital for the working-age population, at least for literacy.

Incremental Earnings Approach

The output of the education sector can be valued using information on how education affects earnings, either by looking at incremental earnings as a function of additional years of school completed or by looking at incremental earnings as a function of improvements in test scores. The prime example of the former approach, also called the present value or lifetime income approach in the literature, is the work of Jorgenson and Fraumeni (1989, 1992). Studies that have estimated the relationship between test scores and earnings include Murnane et al. (1995); Currie and Thomas (1999); and Neal and Johnson (1996).

The Jorgenson and Fraumeni approach to education accounts treats the output of the education sector—years of schooling—as an investment in human capital that is analogous to investments in physical capital such as plant and equipment. Education is regarded as an investment because benefits accrue to an educated individual over a lifetime of activities in the same way that income is generated from the use of physical capital over the lifetime of an asset. This approach explicitly recognizes the gestation period between the application of educational inputs and the realization of benefits. Investment in education is valued as the additional lifetime income from completing another year of school. The theory of investment hypothesizes that the amount that someone will be willing to pay for an investment good is determined by the present discounted value of the future flow of income he or she garners from that good. Accordingly, assuming no externalities, the value of an increment to education will reflect the present discounted value of the additional lifetime income associated with that education. The method used by Jorgenson and Fraumeni (1992) is to compare pairs of demographic groups of the same sex, age, and completed years of schooling, one group enrolled in school and the other not. The present discounted value of the flow of current and expected future income for each pair of groups is compared. The additional lifetime income earned by the group enrolled in school is assumed to arise from the additional year of schooling about to be completed and is used as a measure of the value of the investment in education.[17]

---

[17]It is assumed that enrolled individuals complete the year of school in which they are enrolled. Although theoretically one should be able to link enrollment with years of school completed to form a flow-stock relationship, the two sets of data do not tell a consistent story. A symptom of the apparent problems linking the two data sets is the "rose-colored glasses" effect: as individuals get older some significant portion of them report they have completed more years of education, even though they have not enrolled in school over the relevant time period.

In the Jorgenson and Fraumeni approach, lifetime income includes both market and nonmarket components. Given the substantial amount of time that individuals spend outside of the market, even excluding sleep, the value of nonmarket activities can be several times larger than the value of market activities. Jorgenson and Fraumeni make the strong assumption that education increases productivity in all nonmarket activities, except for the ten hours a day assumed to be devoted to self-maintenance, by the same amount that it increases market productivity. They refer to the assigned value of the time spent in nonmarket activities as nonmarket income. Assigning a uniform increase in value to a full fourteen hours per day of educated individuals' time is somewhat controversial. An alternative would be to assume that, beyond its effect on market productivity, education raises the value only of nonmarket time in which individuals are engaged in productive activity, thereby excluding time devoted to leisure. One also might want to allow for the possibility that the effect of education on nonmarket productivity varies depending on the task involved; as discussed elsewhere in this report, there are certain tasks for which the productivity advantage enjoyed by more educated individuals seems likely to be relatively small. Other researchers who have estimated the nonmarket return to education report substantially smaller figures than do Jorgenson and Fraumeni.[18]

The studies that have estimated incremental earnings as a function of improvements in test scores focus on annual market earnings, rather than lifetime market and nonmarket income. Murnane et al. (1995), for example, estimate that male high school seniors who scored one standard deviation higher on the basic math achievement test in 1980 earned 7.7 percent higher earnings 6 years later, based on data from the High School and Beyond Survey; the comparable figure for females was 10.9 percent. Because this study also controls for students' eventual educational attainment, any effect of cognitive ability as measured by test scores on educational attainment is not counted as a gain from higher test scores.

Currie and Thomas (1999) use the British National Child Development Study to examine the relationship between math and reading test scores at age 7 and earnings at age 33. Estimating a multiple regression of earnings on both test score variables, they find that students who score in the upper quartile of the reading exam earn 20 percent more than students who score in the lower quartile of that exam; similarly, students in the top quartile of the math exam earn another 19 percent more than those in the bottom quartile of that exam. Assuming normality, the average student in the top quartile scores about 2.5 standard deviations higher than the average student in the bottom quartile, so their results imply that a 1.0 standard deviation increase in reading test performance is associated with 8.0

[18]Fraumeni (2000) compares the estimated value of nonmarket activities reported by Eisner (1989), Ironmonger (1996), Landefeld and Howell (1997), and Jorgenson and Fraumeni (1992), and decomposes the Jorgenson and Fraumeni estimate into its various components.

percent higher earnings, while a 1.0 percent standard deviation increase in the math test is associated with 7.6 percent higher earnings. Neal and Johnson (1996) use the National Longitudinal Survey of Youth to estimate the effect of students' scores on the Armed Forces Qualification Test (AFQT) taken at age 15-18 (adjusted for age when the test was taken) on their earnings at age 26-29. They find that a 1.0 standard deviation increase in scores is associated with about 20 percent higher earnings for both men and women. Based on these three studies, a plausible assumption is that a 1.0 standard deviation increase in either math or reading scores is associated with about 8 percent higher earnings.

Of course, noncognitive skills imparted through schooling are also an important component of human capital (see Heckman and Rubinstein, 2001). Unfortunately, measures of noncognitive abilities lag behind available measures of cognitive abilities. In principle, however, the Garrison and Krueger approach could be extended to include noncognitive abilities. If cognitive and noncognitive abilities are correlated, then to some extent the cognitive test measures will reflect noncognitive skills.[19]

Jorgenson and Fraumeni essentially look at the earnings gains due to additional years of schooling, whereas the other researchers discussed above look at earnings gains due to achievement test score gains. Hansen et al. (2004) provide something of a potential bridge between the Jorgenson and Fraumeni method and this approach, as they estimate the effects of additional years of schooling on cognitive test scores. They find that a 1-year increase in schooling is associated with a 0.17 standard deviation increase in the average person's score on the Armed Forces Qualifying Test.

Housing Value Approach

Hedonic housing price models provide another possible technique for estimating the output of the education sector, or at least the output of public primary and secondary education. This approach uses differences in housing prices across areas to estimate the value that parents place on school quality, as indicated by test scores. An often-cited study in this genre, Black (1999) exploits the natural experiment created by different school catchment areas, looking at differences in the values of otherwise similar houses on either side of the border between school zones with different test scores. This approach provides an estimate of the latent price placed on student achievement in the housing market. The close proximity of the houses helps to control for unobserved variables. Her results indicate that parents are willing to pay 2.5 percent more for a house in a district with test scores that are 5 percent higher. Unfortunately, she does not have data on the

[19]See Bowles et al. (2001) for evidence that individual characteristics unrelated to cognitive ability—and perhaps unrelated to productivity—are related to earnings.

inter-student standard deviation of scores, which would be needed as a bridge to scale the results comparably to the NAEP or IALS data.

Kane et al. (2003) follow Black's regression discontinuity design to examine the effect of school test scores on property values in Mecklenberg County (which includes Charlotte), North Carolina, an area encompassing 304,000 real estate parcels and 640,000 people. They have access to data on both the school-level and student-level standard deviation of test scores. In this study, housing prices increase by 18-25 percent for a 1.0 standard deviation increase in test scores (where the standard deviation refers to student-level test score data). At the school level (i.e., using the school-average test score), a 1.0 standard deviation increase in test scores implies a 4-5 percent increase in housing prices. Kane et al. conclude that their estimate of the value parents place on educational attainment is slightly greater than Black's, if put on comparable footing. Other work by Bogart and Cromwell (1997) and Weimer and Wolkoff (2001) also documents a positive relationship between housing prices and school performance.

Estimates such as these could be used to identify the implicit market value of improvements in test scores. One potential advantage of estimates derived from the hedonic housing market literature vis-à-vis the earnings literature is that housing prices presumably take into account the nonwork value of improvements in human capital, or at least that portion of the nonwork value for which parents are willing to pay on behalf of their children (in grades K-12). For example, acquisition of human capital could help individuals to become better parents or better consumers, and these characteristics might be things for which parents would be willing to pay, meaning that their value would be reflected in higher house prices in school districts that produced more of those attributes.

Use of the hedonic approach in an education account should proceed with caution and, certainly, more basic research is needed (the experimental context of satellite accounting should facilitate this). In this literature, which relies on data sources that are fairly limited, unobserved variables appear to be a significant concern. Specification of the functional form and variables to be included in the hedonic regression also may raise questions.[20] As a result, it is difficult to judge how reliable estimates from such studies are relative to, say, those from the incremental earnings studies.

As noted above, the standardized test scores used to measure school district quality reflect only a subset of the valuable attributes fostered by schools. To the extent that different schools have different emphases, test scores may capture their students' relative accomplishments imperfectly. To illustrate, students at a vocational high school might score relatively poorly on standardized tests, but

[20]Specification and identification issues associated with the application of hedonic regression methods are discussed in detail in National Research Council (2002a).

still leave school with highly marketable practical skills that have considerable market value. In addition, the increasing use of test scores to evaluate schools and school administrators could change their meaning in the future. To the extent that schools respond to legislated performance standards by focusing on those areas of learning covered by the tests used in their implementation, at the expense of other important areas that are not covered, the historical association between test scores and human capital may well be altered.

Even accepting test scores as valid measures of human capital, there may be problems with the way that test score information has been used in housing value studies. Parents should be interested in the value added by the school system—in this context, the schools' ability to raise their children's test scores above what they would have been. In practice, researchers have used differences in average test scores across districts as the measure of school district quality. But if average test scores depend on factors such as the education of parents in the district, and these factors are not adequately controlled for, then the estimates could be biased. Neighborhoods with better schools also may have other desirable attributes that contribute to students' accomplishments. At the individual family level, parents who are willing to pay the higher prices associated with homes in higher-test-score school district may spend more time in at-home child education and transmit to their children a greater sense of the importance of education. Black (1999) was particularly concerned about this issue, but her approach of looking at houses very close to school district boundaries where there is discrete variation in school quality should overcome much of the omitted variable problem.

We have argued that, in any experimental satellite account, output should be measured independently of inputs, if at all possible. As we note above, without this independent measurement, productivity in the production of the outputs covered by the account cannot be estimated. As productivity is an important source of economic growth, good estimates of productivity trends are important. There are several possible incremental earnings approaches to valuing the output of education, including the lifetime income approach and approaches that estimate the relationship between test scores and earnings. The implicit valuation of test scores (e.g., NAEP and IALS) using differences in house prices across school districts provides an attractive alternative approach for measuring the value of the output of schools, when the output is measured by performance on standardized tests. On the input side, information on market inputs usually can be found, but obtaining information on nonmarket time inputs and estimating an appropriate wage for valuing those time inputs may be difficult. Once measures of the relevant inputs have been constructed, it is straightforward to create an aggregate input index. In principle, the value of noncognitive skills to which education may contribute could be measured similarly to the value of cognitive skills, although we are not aware of attempts to do so.

Recommendation 5.5: The value of education output should be measured independently of the cost of inputs. Output should be measured using an incremental earnings approach or a housing value approach.

OTHER ISSUES

In the framework outlined in this chapter, there likely will be a substantial "profit" in the education account. This is in part because investment in education may, in fact, deliver a high private return compared with other investments and in part because some of the costs of education are unlikely to be fully accounted for in the education account.

Concerning the former, education may generate a high return because capital constraints prevent workers from investing sufficiently in education (see Ellwood and Kane, 2000; Carneiro and Heckman, 2002), because devoting time to education is viewed as difficult or unpleasant by a substantial number of potential students, or because, for any individual, education is a risky investment and part of the return is a return to risk.[21] In addition, the return could be high because students apply high personal discount rates or underestimate the return to education. And it is possible for the market for human capital to be in temporary disequilibrium because supply can adjust only slowly over time to demand shocks. These concerns, of course, apply to many kinds of markets.

Concerning the latter, investment in children of preschool age is, to some extent, a part of the cost of education that is not measured by the education accounts. In addition, home production necessary to students' school attendance—including activities that range from nurturing and psychological support to preparing school lunches—is another input that is not fully reflected in the education accounts. As with other accounts, it is important to be aware of the reasons for a surplus (or deficit) of output over inputs before making decisions based on the accounts.

[21]In principle, the opportunity cost of time spent in school would reflect the disamenity costs associated with schooling, but this has not been adequately measured.

6

Health

Health has improved dramatically since the industrial revolution. This improvement has resulted from changes in work, life-style, and environmental factors; from advances in public health and medical care; and from increases in wealth, which allow individuals to spend more on goods such as food and shelter that contribute to health. Nordhaus (2003, p. 9) concludes that "accounting for improvements in the health status of the population would make a substantial difference to our measures of economic welfare over the twentieth century in the United States."

This chapter considers the potential for measuring the population's health, and changes in health, within an accounting framework. A population's health is reflected by the average length of its members' lives, and by the quality of those years, as affected by incidence and severity of disease. In contrast to medical care, health itself cannot be purchased directly and, therefore, is not measured in the national income and product accounts (NIPAs). Moreover, with health, there is nothing close to a market equivalent to help us answer valuation questions, so one must turn to other methods.

A fully developed health account would enable researchers to estimate the effect of income and the flow of many other inputs on the stock of health and the value of changes in it. Measuring health is also an important prerequisite for better estimating productivity in medical care. Through these mechanisms, health accounts would facilitate more credible assessment of the desirability of policies that affect provision of health-related goods and services.

Medical care involves substantial expenditures in markets and, to that extent, it is measured in the NIPAs. Where the national accounts have particular difficulty is in decomposing medical spending increases into price and real output. To put it another way, it can be difficult to differentiate among price, quantity, and quality changes. For example, a change in the observed price of treating a disease may reflect a change in the price of unchanged treatment inputs, a change in the amount of inputs (e.g., a surgeon's time) required, the development and use of new drugs or procedures that alter outcomes (e.g., survival rates, patient quality of life), or simultaneous changes in more than one of these factors.

To measure changes in real output, changing quality must be taken into account. For indexes constructed from quantity data, quality adjustments can be made directly. If a new drug is introduced that is equally effective as the old one but only has to be taken once a week instead of every day, the quality (and, in a real sense, the per unit quantity) has increased. If the account is built from price and expenditure data, as in most NIPA components, quality adjustment is made by deflating prices. In the drug example, if the observed unit price increases by anything less than seven-fold, the real price (as dictated by how much has to be spent on the drug per week) actually decreases. Whether a direct quantity-based index or a price-deflated quantity index can be constructed depends on data availability.

This quality measurement problem is prevalent in many industries, not just medical care. When a car costs more because the brakes are better, or when airline tickets cost less because the quality of service is poorer, the change in cost should be attributed to the change in real output, not a change in price. For some manufactured goods, it is possible to estimate the portion of an observed price change attributable to a change in quality in a relatively straightforward way—using direct measurement or hedonic adjustment techniques. Indeed, the Bureau of Labor Statistics (BLS) does just this for a number of goods in construction of its price indexes.[1] The hedonic approach can be problematic for the medical care case. Because most people have health insurance, and because patients are not completely in charge of what they buy, it is not clear that observed prices reflect willingness to pay for particular medical sevices. Since there is not as yet a direct measure of health, or of the contribution of medical care to health, BLS is not able to estimate the productivity of medical care in the usual way. As a result, it is widely believed that some of what is captured as growth in the price of medical care in fact reflects improvements in the quality of care, and that medical care inflation is therefore overstated (see Berndt et al., 2000; Boskin et al., 1996; U.S. Bureau of Labor Statistics, 1997; Shapiro and Wilcox, 1996). Better estimates

[1]Hedonic models are regression equations designed to capture the relationship between a good's characteristics and its price. The estimated coefficients are used to decompose observed changes in price into a portion attributable to changes in items' characteristics and a portion that represents true price change. See National Research Council (2002a: Ch. 4) for a discussion of hedonic methods and National Research Council (2002a: Ch. 6) for recommendations specifically on the pricing of medical care.

of quality change and health output will enable construction of improved measures of medical care productivity.[2]

Finally, health is important to measure because it is an input to other outputs. Healthier people earn more than less healthy people and stay in the labor force longer (see Costa, 2003; Case et al., 2002). In understanding the sources of growth in national income, we need to account for changes in health. In all these ways, a more systematic monitoring of the population's health states, and the factors that have an impact on those states, can advance research and policy.

Recommendation 6.1: A health satellite account should be produced by the Bureau of Economic Analysis in collaboration with the Centers for Medicare and Medicaid Services of the U.S. Department of Health and Human Services.

In this chapter we discuss a framework for developing health accounts and highlight the central choices to be confronted and the work to be done before such accounts can become a reality. While answers are not provided for every possible question, we illustrate what choices are possible and how empirical evidence can inform those choices. In the sections below, we discuss some conceptual issues underlying the idea of health accounts and describe the input and output components of a health account. In the process, we attempt to lay out the major research areas that need to be developed to inform production of such accounts.

CONCEPTUAL FRAMEWORK

As with several of the other areas of nonmarket activity we have examined, we recommend the formation of a satellite account for health—one that does not replace the current NIPAs or components therein, but exists alongside them. In addition to augmenting the picture provided by the NIPAs, development of a satellite health account could uncover approaches and encourage the development of data that improve the way medical care is measured in the conventional accounts.

As discussed earlier in this report, the key to national income accounting is the delineation of inputs and outputs. The total value of output reflects the dollar value of sales on the production side and the dollar value of payments to factors of production on the income side. Since nearly all transactions captured in the NIPAs are market transactions, prices and quantities exist for most items. Money paid by one person to purchase a good or service becomes realized as income to another so, aside from issues relating to measurement of the return capital, the total value of output should be equal to the total value of the inputs used in its production.

---

[2]BLS, adopting recommendations from the National Research Council (2002a), is pursuing new experimental health care price indexes. For a full description of how BLS calculates the medical care component of the Consumer Price Index (CPI), see http://www.bls.gov/cpi/cpifact4.htm.

Because one cannot buy health in a market and because many health science innovations are non-rival in character, there is no guarantee that the value of incremental improvements in health will be equal to their cost. The value of health output can be greater (or less) than the sum of the input values; in a growth context, the change in the value of output can differ from the change in the value of the inputs. To illustrate, imagine that people would be willing to pay $100,000 (in present value) to live a year longer than they currently expect to live. With existing knowledge, no one knows how to guarantee that outcome; hence, their desires go unfulfilled. Now suppose that a new drug is developed that guarantees 1 (and only 1) additional year of life, at a cost of only $1,000 per person, including research and development, manufacturing, and distribution. After the drug is developed, a measure of output based on the cost of the inputs used in its production will increase by much less than the output as valued by consumers. Since consumers are constrained from buying more (years of life), the marginal value of health improvement may well exceed the cost of attaining it.[3]

The possibility of this sort of discrepancy is one of the fundamental reasons for constructing health accounts. A traditional type of measure—such as gross domestic product (GDP)—would include only the money spent improving health. In our example, the contribution of the health sector to GDP would increase by $1,000 per person. In the "knowledge sector," however, at small cost, something was learned that had a large value. If health knowledge is included in the health account, the correct inference is that the health sector has become more productive. Measuring the output "health" and the input of dollars (and time) spent to improve health, including dollars spent on research and development, allows calculation of how much more productive.

Market-Oriented Approaches

The market-transacted goods and services relating to medical care appear in various household consumption, government, and investment components of GDP. Examples are direct medical care spending, public medical care, and pharmaceutical research and development. Since 1964, the Department of Health and Human Services has published National Health Accounts that include time-series data on national health expenditures. The health accounts seek to "identify all goods and services that can be characterized as relating to health care in the nation, and determine the amount of money used for the purchase of these goods and services . . ." (Rice et al., 1982).

---

[3]The point here is that the limited (and discrete) effectiveness of the pill eliminates the possibility of purchasing pills up to the point where the "the marginal value of health" equals the price of the pill. If additional pills could add additional years of life at the same cost, consumers might buy many more, until the value of the additional year of life was as low as $1,000. In this case, valuing the pills at their $1,000 cost would be as good as similar valuations for shirts or shoes.

Medical care and health research expenditures account for a large and growing percentage of GDP (14.9 percent in 2002, up from about 9 percent in 1981 and from just over 5 percent in 1961). Table 6-1 shows the sources and uses of those expenditures. Note that the data are, by and large, organized by payer group or by type of institution making or receiving payments.

From a national accounts perspective, the essential problem with health expenditure data, as currently collected, is that they do not provide price and quantity information about anything that might reasonably be considered an output. The expenditure data answer only the questions, "Where does the money come from?" and "Whom does it go to?" They do not answer the central question, "What does it buy?" (Triplett, 2001, p. 3). In this sense, the current medical care accounts are undeveloped, even by the market-oriented NIPA standards.

What is the output of medical care? The Eurostat *Handbook* (Triplett, 2001) identifies "completed treatments" as the output of the health care sector of the economy. Ultimately, what people want from medical care is improved health, and one way they pursue this is through treatment of ailments and diseases. Doctors' time, hospital patient days, drugs—things typically treated as outputs—are more accurately viewed as inputs used in the production of treatments. In practice, one would measure this at the disease level—how much has health improved for people with a particular disease. Though research on cost-of-disease or cost-of-treatment approaches is embryonic, prices of these service sets could

TABLE 6-1 The Nation's Health Dollar, 2002 (percent of expenditures)

Sources		Uses	
Private insurance	35	Hospital care	31
Medicare	17	Physician and clinical services	22
Medicaid and SCHIP[a]	16	Prescription drugs	11
Out-of-pocket	14	Nursing home care	7
Other public[b]	12	Program administration and net cost	7
Other private[c]	5	Other spending[d]	22

[a]State Children's Health Insurance Program.

[b]Includes programs such as workers' compensation, public health activity, Department of Defense, Department of Veterans Affairs, Indian Health Service, and state and local hospital subsidies and school health programs.

[c]Includes industrial in-plant, privately funded construction, and nonpatient revenues, including philanthropy.

[d]Includes dental services, other professional services, home health care, durable medical products, over-the-counter medicines and sundries, public health activities, research, and construction.

SOURCE: Centers for Medicare and Medicaid Services, Office of the Actuary, National Health Statistics Group. The website for the Department of Health and Human Services has detailed tables of national health expenditures, by source of funds and type and amount of expenditure (http://www.cms.hhs.gov/statistics/nhe/historical/chart.asp).

in principle be tracked by type, then quality adjusted and weighted appropriately (Triplett, 2001, p. 1). The current lack of pricing by treatment is, to some extent, a data collection issue. As shown in Table 6-1, data are currently collected by institution and not by treatment. Quality adjustment at the treatment level is another difficult issue. Yet research by Cutler et al. (1998), Shapiro and Wilcox (1996), Triplett (1999), and others has been moving in this direction over the past decade, and solutions to difficult productivity and pricing issues are emerging.

The treatment-based cost-of-disease approach is more consistent with the way the rest of the market-oriented NIPAs are designed: "if the medical care sector is the industry, then the treatments are the products the industry produces" (Triplett, 2001, pp. 3-4). While recognition is growing that treatments and outcomes—not time with the doctor, days in the hospital, and so on—are the conceptually relevant units, measurement of the appropriate quantities and prices is, at this point, very incomplete. The Aging-Related Diseases (ARD) Study of the Organisation of Economic Co-operation and Development (OECD) has begun cataloging information on the cost of treatments in a few areas such as heart disease, stroke, and breast cancer (Triplett, 2002). Research on cost-of-disease accounts is progressing in several countries, including the United States, Canada, the United Kingdom, Australia, and the Netherlands. This kind of work begins with a reorganization of data from the national economic accounts into expenditures by international disease classification codes. Much of this work in the United States is being done by the National Center for Health Statistics (NCHS).

A Broader Approach

Research on disease and treatment-cost frameworks has great potential to improve the usefulness of national health care accounts. For a health (not health care) account that would fit into a set of satellite nonmarket accounts, however, one would want to go much further. Specifically, in accord with the rest of this report, we advocate an account that (1) includes both market and nonmarket inputs and outputs, (2) estimates input and output values independently, and (3) defines outputs that are linked to utility as directly as possible.

While the treatment-based approach described above redefines medical sector output in a way that is appropriate for the NIPAs, it does not necessarily advance the objective of measuring inputs and outputs independently. For example, the technique for pricing heart disease treatment may simply sum up the costs of inputs, such as those associated with angioplasty, hospital time, and pharmaceuticals. Most of the data in the ARD Study capture input costs since prices are generally not charged directly on the basis of complete medical treatments. In a prospective payment system, one might be able to obtain reasonably good cost estimates for most one-time procedures that do not result in complications (e.g., a bypass operation, or the delivery of a baby), but one does not get a quote for, say, full treatment of heart disease in the way one might for repairing a broken transmission on an automobile (Triplett, 2002, p. 9).

More important is the question of how output is defined: utility is generated by improved health, which is a nonmarket good, not by the treatment itself. Unlike, say, a vacation or trip to a restaurant, most people do not subject themselves to medical treatment for the fun of it—quite the contrary. And medical care is not the only, and possibly not even the largest, input to the production of health. There are in fact a large number of factors—both medical and nonmedical, and market and nonmarket—that influence health. In this context, health condition is the output and medical treatment is one of many inputs. It is this concept of health that we are primarily concerned with for nonmarket accounts.

The conceptual framework for this strain of health accounting has its antecedents in the literature on the determinants of health and life expectancy. Just as national health expenditures have changed, so too has the nation's health, perhaps even more dramatically (see Nordhaus, 2003; Murphy and Topel, 2003). By most measures, improvements in the health of the population have outpaced the increase in spending on medical care; since improvements in health are related to a broad range of interrelated factors, however, one cannot say for certain what factors are most important (Cutler and Richardson, 1997; Cutler, 2004).

The literature—which includes McKeown (1976), Fogel, (1986, 2004), Preston (1993), Riley (2001) and many others—attempts to attribute health improvements, typically as measured by life expectancy, to determinants such as medical care, diet and nutrition, and environmental and life-style factors. McKeown (1976), for example, demonstrated that, in England and Wales, declines in mortality between 1900 and 1970 resulted primarily from reductions in the incidence of infectious diseases such as tuberculosis and smallpox. Furthermore, the bulk of this health improvement occurred prior to the development of effective medical therapies. He attributes much of the reduction in disease prevalence to improvements in basic living standards, particularly nutrition. Fogel also focused on diet, using height and body mass as indicators of nutritional status. His research suggests that nutrition played an especially large role during the eighteenth and nineteenth centuries, explaining perhaps 90 percent of the decline in mortality for England and France in that period. For the United States, where diets had significantly improved by the turn of the century, Preston and Haines (1991) argues that improvements in public health—ranging from better hygiene to swamp drainage to mosquito control—were central to reducing waterborne diseases, which reduced mortality rates both directly and indirectly through nutritional consequences. Cutler (2004) argues that major advances in medicine constitute the dominant explanation for improved health in the United States since the mid-twentieth century.

Indeed, a primary reason for developing an experimental health account is to add rigor to research aimed at attributing improvements in health to specific types of expenditures, behavioral changes, and increases in basic medical science

TABLE 6-2 Health Inputs and Outputs, Market and Nonmarket

Inputs	Outputs
Medical care	Health status
Market labor/capital	Longevity
Volunteer labor	Quality of life
Time invested in individual's own health	
Other consumption items	
Research and development	
Quality of environment	

knowledge. Statistics on causes of death are extremely helpful to policy; consider, for example the success of campaigns to increase seat belt use. Still, mortality rates alone present an incomplete picture of a population's health, and the analyses noted above focus on very long-term trends. An ideal, and more sensitive, health metric—and one that would be most useful to health policy—would incorporate both fatal and nonfatal health outcomes. For example, in planning how to target public health expenditures, it would be extremely helpful to know with accuracy the extent to which smoking, obesity, and other factors contribute to disease burden.[4] It is this concern that has led to development of measures of health status such as quality-adjusted life years (QALYs). We return to these related measures in our discussion of outputs below.

Forming nonmarket health accounts begins with identifying the key inputs and outputs. One possible structure is suggested in Table 6-2. Many of the inputs and outputs of the health sector are stocks. An early developer of the framework for thinking about health in this way was Michael Grossman (1972). His model portrays health as "a durable capital stock that yields an output of healthy time. . . . Individuals inherit an initial amount of this stock that depreciates with age and can be increased by investment" (p. 1). This interpretation is most obvious with health, but it also applies to other nonmarket areas—the stock of research and development and the quality of the environment, for example. And the value of these stocks may be quite large: as we discuss below, health capital has been estimated to be roughly $7 million per person.

In practice, though, we frequently want to measure changes in health capital, rather than the level. For example, to determine the productivity of the medical system, one would want to know how much is spent on medical care in a year and compare that to the value of the improvement in health resulting from that intervention. For this question, measuring the endowment of health is less important

[4]The major ideas underlying burden-of-disease measures can be found in Murray and Lopez (1996).

than determining how health changes. This is fortunate because empirical work generally has estimated the value of incremental health improvement, not the value of the health endowment itself.

We envision a data system designed to monitor changes in health—as measured by life expectancy, disease prevalence, and activity impairments—and as many of the determinants of that change as possible. Measuring productivity in the health sector requires keeping track of the flow of inputs and isolating the contribution of each to better health. If a policy change extends a treatment using existing medical therapies to more people in an afflicted group, we would like to know how that policy affects health outcomes. It is essential, within this framework, to track separately the various inputs to health. When the relationships among the health inputs, their interactions, and the health output are not fully understood, it makes little sense to try to aggregate the inputs. A satellite health account should, at least initially, be oriented toward providing data that could be used to estimate productivity at the disease level.

Clearly, identifying the effect of medical care on health, separate from the effects of other inputs, is difficult. But the problem does not need to be solved in order to begin work to construct a health account. The immediate goal is to devise an approach to measuring health and the various inputs to health, leaving questions of causal relationships to others. Because policy demands more complete information about health sector productivity, however, the need for solutions to these questions is the primary reason for pursuing the health accounts. The long-range goal is to be able to identify the influence of such inputs as diet, treatments, and life-style on health status, at least on a disease-by-disease basis. As research reveals new information about these relationships, the value of the data collected in a health account will compound.

MEASURING AND VALUING INPUTS

There are at least six major inputs to the production of health:

- medical care provided in market settings;
- medical care services provided without payment;
- time that individuals invest in their own health;
- consumption of nonmedical goods and services, some of which may improve health and others of which are harmful, and nonmedical technology and safety devices;
- research and development that may lead to improvements in medical technology and knowledge, and
- environmental and "disease state" factors (and shocks).

The first major input—medical care provided in market settings—is readily measurable. When sick, people may visit doctors, undergo tests, and take medications. The national income accounts already measure the monetary flows associated with payments for these goods and services. As noted, however, there are a variety of other inputs to health that also should be measured.

> **Recommendation 6.2:** Market inputs to health such as expenditures on medical care—already measured in the NIPAs—should be included in the health satellite account. The account should go further, however, and measure both the quantity and quality of medical care and include nonmarket inputs relating to time, diet, exercise, and other factors.

The second major health input—care service provided without payment—is missed in the national accounts. There are two types of noncompensated care services—volunteer services and services provided by family members. Many hospitals utilize volunteer labor, for example, to perform tasks ranging from fund raising to providing ambulance service. The more important category of noncompensated services is that of family members who provide care for sick or injured relatives. For an elderly person with infirmities who is cared for by a spouse or child, no monetary transaction is involved, and so the services are not reflected in the national income accounts. This exclusion is particularly important in the case of long-term care services. Recent estimates suggest that the value of nonmonetary long-term care services may be even greater than the value of market-provided services. LaPlante et al., (2002)—using data from the National Health Interview Survey on Disability, a nationally representative household survey conducted between 1994 and 1997—show that more people receive unpaid personal assistance services than paid services, and that the average weekly amount of unpaid help per noninstitutionalized adult is also higher.

It is easy to see the relevance of measuring unpaid care time to health care policy. For example, apparent cost savings associated with recent reductions in the length of hospital stays have been partly offset by an increased nonmarket burden on families who have to care for patients who are discharged "quicker and sicker" (Pamuk et al., 1998). In such a case, policy makers should be aware of both market and nonmarket costs.

Until recently, the data available to measure volunteer and family time devoted to health care activities have been limited. For the past several years, a supplement to the Current Population Survey has collected information on volunteerism, which it defines as persons who do unpaid work for or through an organization. There are real questions, however, about how accurately people can report time devoted to such activities over a period as long as a year and the coding of the type of volunteer work performed is very broad. The information on time use being collected in the American Time Use Survey (ATUS) should be both more accurate and more detailed. Many of the relevant activities will be reported within the "caring for and helping household members" top-level category; some entries in the "caring for and helping non-household members" category also will be relevant.

> **Recommendation 6.3:** Unpaid time spent providing health-related services—primarily volunteer work and intrafamily care—should be measured and included in the health satellite account. The amount of time spent in these activities that improve or maintain health should be calculated using data from the new BLS time-use survey and other relevant sources.

As with any nonmarket transaction, there is a question about what wage to use in valuing these unpaid services. Consider a middle-aged woman who earns $25 per hour in the labor force and also spends time caring for elderly parents. Should the implicit price of the caring services be $25 an hour (or some function thereof), or $10 per hour, the amount that such services might (hypothetically) cost in the market? Our general answer, as discussed earlier in this report, is to value time devoted to activities that someone else could have been hired to perform at the market replacement cost. The example given earlier was that time spent by a homeowner putting a roof on the house should be valued at a roofer's wage, adjusted for the relative productivity of the homeowner as compared to the professional roofer. For a person taking care of elderly parents, the market alternative is to hire a caretaker at the prevailing wage for that occupation.

> **Recommendation 6.4:** Unpaid time spent providing health-related services should be valued based on a replacement labor cost approach, adjusted for productivity and quality differences between the services provided by the market and the unpaid provider.

In other applications discussed in this report, there is reason to think of the replacement wage as an upper bound for the purpose of valuing nonmarket time, since market service providers often offer more specialized skills than those possessed by a home producer. This may not be the case, however, for health care, elder care, or child care, where family members may offer the more nurturing or caring option.

It is interesting that, in many cases, individuals choose to care for family members even when their market wage is above that paid to hired caregivers—that is, even when the market value of their time input is less than they could have earned by devoting the time to labor-market activity. As in earlier applications, we interpret the difference between the individual's market wage and the replacement cost of the caregiver services provided to a family member as an estimate of the consumption value received from supplying the service personally rather than through the market. We do not recommend, however, that this consumption value be included in the output column of the health account. The satisfaction associated with caring for a family member does not contribute directly to the production of health, which is what the health account is trying to measure.

The third major input to the production of health is time that individuals invest in their own health. People exercise to prevent heart disease, sleep to refresh their bodies, and spend time out of the labor force to recover from illness.

These time costs are investments, just as is medical care. Unlike the case of middle-aged people taking care of their parents, there are no markets in which comparable personal services can be bought. People cannot pay someone else to sleep or exercise for them, the way that they could pay people to care for their parents. The various own-time inputs that contribute to health fall into the class of nonmarket goods that do not satisfy the third-person criterion, which eliminates the possibility of using a market-substitutes-based valuation method. In valuing these personal time investments, therefore, there is no real alternative but to price them at the individual's own time cost.

Recommendation 6.5: Time spent by individuals in activities that improve or maintain their own health should be included in the account and valued at the opportunity cost, modified to net out the consumption or enjoyment value of the activity.

Much of what should be counted would be captured in the "sports, exercise, and recreation" top-level category of the ATUS, but there could be activities in other categories that should be counted as well.

Opportunity cost, presumably some function of an individual's own market wage, is conceptually the right approach for many activities. There may be some exceptions. For example, time spent recuperating from illness or injury really does not have as its opportunity cost the person's wage rate: if a person is ill, she or he will not be very productive in any activity except recuperating. Someone in recovery may not actually be foregoing a wage since, during that period, she or he is probably not worth much on the job.

There are also some practical complications associated with opportunity cost methods. For example, there is the complication of valuing the time of individuals not in the labor force (or unemployed). Perhaps more importantly, as has been discussed earlier in the report, time is not a homogenous good, implying that the market value of hours spent caring for family members need not equal an individual's marginal or average hourly rate, though it is likely that opportunity cost will be positively correlated with the wage rate.

More difficult yet is the question of what activities to include in the account as time inputs to the production of one's own health. There are valid arguments for not including (or at least not attempting to value) all of the things people may do that affect their health. Valuing sleep is particularly problematic. Sleep represents a large chunk of time, but it is undertaken for a number of reasons. Therefore, even if one could calculate the value of sleep, one would want to include only its marginal contribution to health (such as that linked to recuperation from illness). There is a meaningful tradeoff between market time and the last hour of sleep, but does the same tradeoff exist for the first hour of sleep? Put differently, what would happen to a worker's hourly wage if he or she attempted to deliver 24 hours of market work with zero sleep? Undoubtedly, it would be lower than the hourly wage observed for a standard workday, and it would quickly fall to zero as sleeplessness drove productivity to zero. In this sense, the first hour of sleep is worth less than the last hour because it cannot be productively converted into market time. But the first hour is worth more to the individual's health and general well-being and presumably will have a larger effect on productivity while awake. Given the inherent difficulties here, and until it can be shown that changing sleep patterns have had a significant effect on health, it would not be sensible to assign a value to sleep time in a health account. It might be useful, however, to include data on hours of sleep from time-use studies, since it might have health implications.

Recommendation 6.5 specifies that the enjoyment derived from health-enhancing activities should be netted out of the time-unit valuation included in the account. The netting-out rule for valuation of own time devoted to health-enhancing activities comes into play for the many forms of exercise from which individuals derive pleasure. For many people, an evening walk, weekend kayaking expedition, or recreational soccer game is a pleasure as well as an investment in cardiovascular health. The case can certainly be made that such nonmarket recreation has value that should be accounted for somewhere, perhaps in a "recreation account" developed as another component of a comprehensive accounting of nonmarket activities. But for the health account, specifically, one wants to net out this recreation value because it does not relate directly to health. Whether or not one wishes to encourage development of a recreation account is a separate issue.

Consumption of other goods and services is the fourth major item affecting health. Cigarette and alcohol use, intake of fatty foods, and inclusion of fruits and vegetables in the diet all affect health. By definition, spending on these consumption items already is measured in the market. Some goods (e.g., airbags) can be expected to improve health and thus have a positive input value; other items (e.g., cigarettes) can be expected to harm health and thus have a negative input value. The prices of these items reflect their current consumption value in addition to whatever consumers are willing to pay for their perceived effects on future health. Because they occur only in the future—and sometimes the distant future—both the benefits of healthy behavior and the costs of unhealthy behavior will be discounted by individuals at the time of consumption. Moreover, health-linked behavior and consumption affect only the probability of particular health outcomes. Both the discounting and the probabilistic nature of any health effects associated with consumption patterns would be accounted for in measures of the stock of health—for example, in data relating change in prevalence of smoking to changes in a population's health.

Data on such dietary habits as smoking, drinking, and fat and cholesterol intake should be collated and included in the health satellite account, at least in an auxiliary way, even if valuation questions remain unsolved. As noted above, these data will be useful to researchers attempting to tie health outputs to specific inputs. Information on how consumption of purchased goods and services affects health outcomes (morbidity and mortality)—and specifically how diet and nutrient intake affect health capital—is an essential component on the input side of a health account. Data on products in the nonmedical technology category—products such as seat belts, air bags, and smoke detectors—should also be tracked. Even though the way in which these factors combine with others to affect a population's health status is not yet exactly known, information about them will be useful for research on health determinants and medical care productivity.

The fifth input to health is technology created by research and development (R&D). Research and development generates information and intellectual capital which later may produce flows of services in the form of medical advances and other innovations. Some R&D occurs in the private sector, generally by pharmaceutical and medical device companies. Other R&D is paid for by the public sector and occurs in national laboratories or at universities. In the case of the former, research costs generally are reflected in the prices of the drugs that are sold. The health account should not value this type of R&D separately when it would double count the market component of the account. The benefits from basic science research in a university or nonprofit setting are more likely to be missed in the market accounts and, hence, to need special attention. There are estimates of the flow amount of R&D that can be entered into the health account.

> **Recommendation 6.6:** New technologies created by research and development expenditures should be measured and included in the health accounts.

One research area that is not well covered in the NIPAs is the development of non-rival innovations. Much of this work is likely to take place in nonprofit research settings. Initial work on valuing nonmarket R&D might reasonably focus on such organizations. Additionally, in general, the value of research and development should be capitalized to reflect depreciation and obsolescence.

The sixth input category includes the various environmental factors that affect health. Air pollution and unclean water harm health, while public parks may improve health. Valuation of the environment is a topic that is dealt with elsewhere in this report and more extensively in *Nature's Numbers* (National Research Council, 1999). We thus do not take it up here, other than to say that the environment enters the health production function alongside diet, life-style, and other factors that ultimately could be brought into a health satellite account. In so doing, it should be recognized that only a subset of environmental factors influence health. To take a simple example, air pollution may lead to an increased incidence of asthma and to an obscured view of the Grand Canyon. A health account would be concerned only with the former. More research is needed before the role of these "exogenous" factors—changes in the physical environment or the appearance of new diseases—can be understood. Keeping track of these factors in a set of accounts would be a useful step.

While we see the above-described components of a health account as the most important, there are other factors to keep in mind. For example, the demographic make-up of a population will affect measures of "average health." All else equal, one would expect a younger population to be classified as more healthy than an older one by most measures. The issue of how to separate pure age effects from changes in the disease environment and other shocks requires further study. Ideally, an account would coordinate data useful to researchers who want to estimate the effect of changing demographics on health levels. As the health account is developed, perhaps a demographics module (such as that discussed in Chapter 2) can be integrated. Not only is the demographic information necessary for estimating a population's health stock, it also would be needed for estimating per capita effects (analogous to per capita GDP).

MEASURING AND VALUING HEALTH

Defining the Output

Two outputs are associated with investments in health. The first is the flow of better health that results from the investment. The present discounted value of this flow is termed health capital. People enjoy being healthier just as they enjoy consuming better food or nicer clothing.[5] The second output of investments in health is the additional income that a healthier population generates. A complete set of health accounts would include the present value of expected future earnings, perhaps augmented to include the value of expected future home production, that results from changed health. This is analogous to the way human capital is valued in the labor economics literature.

> **Recommendation 6.7:** In the health satellite account, output should be measured independently of inputs (and the two need not be equal). Changes in both components of output—the consumption flow of good health and the additional (or reduced) income that a healthier (or less healthy) population earns—should be measured.

Health events, such as the contraction of a disease, affect the future consumption flow of health. These events, as well as interventions in response to those events, impact individuals' quality of life across future periods, often in unpredictable ways. Money spent on a bypass surgery operation often will improve quality of life, and possibly reduce future mortality rates. Likewise, present levels of health are linked to past actions—the production of health is an intertemporal process tied to past levels of health care, diet, environmental and other factors (Triplett, 2002, p. 2).

The second output, that relating to the additional income that a healthier population earns, is most starkly observed in poorer countries, where many people have substantially impaired productivity because of poor levels of health (Strauss

[5]Michael Grossman (1972) cites Jeremy Bentham as including health in one of 15 "simple pleasures."

and Thomas, 1998). But the link between health and income exists in developed countries as well. For example, people who are depressed are less likely to finish their education, more likely to drop out of the labor force, and less productive at work than are people who are not depressed (see Sturm et al., 1999, Leroux et al., 2003). Similarly, older individuals who suffer adverse health shocks are more likely to drop out of the labor force than are people who do not suffer such shocks (Smith, 2004). A health account thus would include the present value of the expected flow of future earnings, as in the human capital account.

The economic return to better health can be measured analogously to the economic return to education (see Chapter 5). Wage equations can be estimated to determine how health affects total income, hours of work, and wages. There are a host of econometric issues in specifying these equations: measurement of health is difficult; health may be endogenous to income; other factors such as pollution or smoking may influence both health and income, and so on (Farrell and Fuchs, 1982). The armamentarium of modern econometrics can be brought to bear on these problems, however, and it is reasonable to expect good results.

The novel challenge in constructing health accounts is the measurement of health. Without a measure of health, one cannot monitor and value the central outputs of the medical system. Because health is a multidimensional concept, encompassing length and quality of life and involving both mental and physical health, it is very difficult to measure. Most of the research to date has sought a method of summarizing information on diverse health states using a common metric and then valuing a common increment on that scale. The scale typically is called a quality-of-life scale, and the resulting index of the stock of health a quality-adjusted life expectancy measure (for a detailed discussion, see Weinstein et al., 1996).

Imagine ranking all medical conditions on a scale ranging from 0 to 1, where 0 is death and 1 is perfect health. If no state is worse than death, every health state can be placed on that interval. A person has a quality of life each year, which may vary over time. Now suppose that we know the value of some particular increment on the common scale—for example, the value of perfect health relative to death. Then, the value of an individual's health state can be given as the product of the present value of her or his quality of life relative to perfect health, times the value of a year in perfect health. Formally, the value of health is defined as:

$$\text{Value of health} = V\sum_{t=0}^{T} \delta^t q_t, \quad (6.1)$$

where V is the value of perfect health; δ^t is the applicable discount rate and q_t is an individual's quality of life for year t.

This framework is common in the literature, but it should be noted that it is not innocuous. It implies, for example, that similar quality-of-life improvements are worth the same amount at all ages; that people are not risk averse; that the utility from health in one period is independent of health status in other periods; and that utility from longevity is independent of the utility from health-state level (for a discussion of the assumptions underlying various health valuation metrics, see Hammitt, 2003; Eeckhoudt and Hammitt, 2001). Still, some such structural representation, perhaps with the introduction of age-specific adjustments, is the way most researchers have made progress on this difficult issue.

There are important distinctions between how health output would be valued in conjunction with the frameworks just discussed and how market goods and services are valued in the NIPAs. There are often gaps between individuals' willingness to pay for a good or service and the cost of providing that good or service. Consider a serendipitous discovery of a new technique for curing a terrible disease. Suppose further that this technique costs little or nothing to implement. In the health accounting frameworks discussed above, this discovery would add significant value because it enhances health. Put another way, value is high because willingness to pay is high. In contrast, in a NIPA framework, this discovery would have little or no value because little or no resources are required in order to enjoy it. Things of value that require no resources have no value in a NIPA framework.

The distinction between what one would be willing to pay for a good or service and what one has to pay is ubiquitous in the valuing of nonmarket transactions. Measurement procedures that use willingness to pay for nonmarket activities are not compatible with how market activity is measured, and overstate the value of nonmarket activities relative to conventional measures of market activities.

One can envision that, for many health policy purposes, it would be helpful to use a willingness-to-pay metric to value health output. For other applications, it might be conceptually preferable to use more conventional valuations even though, in many instances, it may be impossible to estimate anything that would reasonably serve as an analog to a market price. Sometimes the nature of available data may not leave much choice in the matter; either way, though, health accounts must be transparent with respect to how increments to health are valued. Ideally, when the health account produces a valuation that includes consumer surplus, an attempt should be made to decompose total value into surplus and product to allow comparability with the NIPA measurement framework.

Measuring Health Status

The first step in measuring health output is to develop a scale along which different health states can be compared. The objective is to develop measures that allow changes in a population's expected quantity and quality of life to be estimated. Such measures would be capable, for example, of reflecting an average change in the length or quality of life for a population in response to the appearance of a new disease, to an effective new treatment for an old disease, or to changes in environmental factors. There is a vibrant and growing literature on the measurement of health states, which we draw from in this section.

The standards for what constitutes good health, as well as expectations regarding length of life, shift over time. Still, the panel believes that health should be viewed in absolute terms. That is, measures of health—such as life expectancy, morbidity, and ability to perform daily tasks without pain or other limiting factors—are cardinal measures. They can and should be measured consistently over time.[6] A health measurement should show improvement in health whenever the distribution of individuals across health outcomes improves, even if expectations about health are rising along with health outcomes. With this in mind, we highlight two approaches to health status measurement.

The first approach—the disease state approach—involves estimating the quality of life for people with different diseases and then multiplying those measures by disease prevalence in the population making adjustments to reflect the percentage of people with multiple conditions. A population's stock of health capital changes with the prevalence of diseases or with changes in the impact of these diseases on human functioning. Disease-state measurement is complicated by the fact that diseases become more or less prevalent over time for reasons that are hard to explain. New diseases appear (e.g., AIDS, SARS); others fade away or are no longer diagnosed (e.g., neuralgia); and various diseases are at different stages of the proliferation or containment cycle in different regions of the world.

A second, related, approach to health measurement—the health-impairment approach—involves asking people about ways that health problems interfere with their lives. Since diseases presumably have effects that interfere with normal activities, this is a more direct way to measure health. As with the disease-state approach, the idea is to quantify changes in the population's health status that result from various health events or from investments in the inputs to health (medical care, time, research, etc.) made in response to those events. One factor that complicates this type of measurement is that characteristics such as age, gender, life-style, motivation, and preferences may modify the impact of an illness on different individuals; these characteristics also influence the efficacy of treatments.

Within these two approaches, there are several ways to confront the need to quantify the relative severity of different conditions. One way is to survey experts, presumably physicians and health researchers, about the extent to which diseases affect quality of life. As with any of the available options, this would entail assigning a numerical score to various disease or impairment states. For example,

[6]This does not mean that the relationship between a given health impairment and utility must remain constant. For example, the disutility of a walking impairment may be less today than it was 200 years ago when there was less infrastructure developed to accommodate people with physical limitations. The weights associated with impairments on the quality-of-life scale may have to be adjusted over time, as might the value of a statistical life.

someone with a walking impairment might be assigned a 0.7, relative to someone in perfect health who would receive a 1.0. It is not clear, though, that experts in disease and treatment necessarily have an advantage in evaluating the relative utilities associated with different impaired conditions or disease states.

A second approach, around which a significant literature has grown, involves asking individuals to assess their current health status against counterfactual alternatives. The most theoretically appealing health state measurements involve explicit utility comparisons. For example, a time tradeoff approach can be used in which a person with, say, arthritis is asked a question of the form: "Suppose you have 10 years to live in your current health state. How many years would you be willing to give up to live without arthritis?" A related approach is a standard gamble. Here, the subject is asked: "Imagine that an operation is developed that will cure your arthritis but, with some probability, you will not survive the operation. What chance of success would you need in order to go ahead with the operation?" These types of questions can be used to construct a utility assessment for arthritis.

In practice, the most common way to value health states is with simple multiple choice questions: "How would you describe your health—excellent, very good, good, fair, or poor?" or with easy-to-use scales: "On a scale of 0 to 100, how would you rate your health currently?" (Torrance, 1987). The answers then can be compared across people.

There is an important difference between these scales and utility-based measures that require a survey respondent to compare his or her health state against that of someone in a reference (healthy) population. In the former, one generally compares health ratings of people with a condition to the ratings of others without the condition; people are not asked to evaluate the counterfactual of living a life different from the ones they have. In the latter, people are asked to evaluate their current health state and then to evaluate an alternative with or without a particular condition. One possible advantage of this self-evaluation method is that it controls for other factors that differ across the population. The problem is the speculative nature of such surveys (the definitive volume on subjective evaluation is Kahneman et al., 1999). For instance, the evidence shows that people with an affliction (such as paralysis) report systematically different quality-of-life scores from people without the affliction who are asked to rate what their health would be with the affliction. By comparing survey results that ask people to rate their own health as it is, one may be better able to attribute differences in average satisfaction ratings to a specific condition.[7] If the other important factors influencing self health assessments can be controlled for and embedded, estimates of

---

[7]There are still complications. For example, while it is easier for respondents to provide answers using a linear analog scale, that approach also has known biases, in that people tend to choose answers closer to the middle than they do with alternative approaches.

the impact of a condition will be unbiased (Cutler and Richardson, 1997:252). To date, there has been little comparison of these approaches; more is needed.

A number of health assessment questionnaires have been developed. The names indicate the degree of experimentation in the field: SF-36 (for short form, 36 items; the longer form was developed in the 1970s); Euroqual-5D (for five domains); the QWB (quality of well-being) survey; YHL (years of healthy life expectancy); and so on. Among these instruments, common dimensions include impairment in functioning due to physical problems, impairment in functioning due to mental health issues, pain and discomfort, and vitality. Surveys may also attempt to track intervention-related factors such as side effects to medication or even the dollar cost of treatments. Questionnaires ask about a range of activities, from those that are directly economic, such as one's ability to work, to more basic indicators of functioning, such as eating, arising, dressing, reaching, and general mobility.[8]

The health assessment surveys are designed to measure health-related quality of life with single numbers that represent the population's preferences for combinations of symptoms relating to mobility, physical activity, and social activity. The scales grade symptoms by the degree to which they affect everyday activities. A lengthy comparative literature is being developed, with adherents of each of these approaches, especially for characterizing health outcomes over time for patients having such serious illnesses as cancer or AIDS.

The federal government, itself a producer of health assessment measures, uses them in policy formation.[9] Every decade, the U.S. Department of Health and Human Services outlines a set of "healthy people" goals: the first was for 1990; subsequent editions have been issued for 2000 and 2010 (see U.S. Department of Health and Human Services, 2000). These decennial volumes provide a thorough overview of work on these approaches—including the YHL approach noted above—and report results for a set of survey measures designed to track the population's health. Among the measures tracked are: self-rated health and recent days of physical health, mental health, and activity limitations. Metrics such as "quality-adjusted life-years" and "disease-adjusted life-years" are already widely used in medical applications to identify unmet health needs, and to guide policies for addressing those needs.

---

[8]Survey instruments designed to estimate the relationship between health impairments and the ability to carry out work often do so both in terms of increased absence and reduced productivity. The Health and Labor Questionnaire, the Work Limitations Questionnaire, and the Work Productivity and Activity Impairment Questionnaire are just a few of these surveys. The results from these instruments are often validated against other measures such as quality-of-life scales, and typically measure productivity in percentage terms against an unaffected group (Reilly et al., 1993).

[9]The Centers for Medicare and Medicaid Services in the Department of Health and Human Services also maintain a set of "national health accounts," but (as described above) these are spending accounts, not health accounts.

Despite its vibrancy, this area of research is quite new, and great care needs to be taken when comparing survey measures of health status and well-being across groups, even within rich countries. Because most measures of health status are subjective, differential reporting norms in different groups may affect quality-of-life estimates for those groups. For example, women may be less inclined to describe themselves in extreme terms than are men. As a result, women may report their health as being more towards the middle of the health distribution than men do, less excellent and less poor. This inclination might lead to the misperception that adverse health conditions affecting women are less serious than those affecting men.

Recent research has suggested two approaches for addressing response comparability problems. The first is to ask questions about hypothetical, researcher-designed health conditions of different groups. These "vignettes" can be used to benchmark responses that individuals in different groups give about their own health (King et al., 1998). For example, in some cases, self-reported health may differ by gender; women offer less extreme responses than do men to certain kinds of identical vignette questions. This kind of information could be used to adjust, at the group level, different response patterns about own health.

A second solution, alluded to earlier, is to make comparisons of health conditions across individuals within a group and then standardize across groups using a common benchmark (Cutler and Richardson, 1997). For example, one could examine the health of women with and without a particular condition and form the health decrement of the condition as the difference between the two. The health decrement for women then can be compared to the health decrement for men. To make an explicit quality-of-life estimate, one might impose, for example, the assumption that people of the same age with no adverse health problems had the same quality of life.

Measures of health should be based on objective observation of health outcomes. The field of health assessment has not settled on one approach to measuring health as superior to the others. At this point, the full range of methods, each of which attempts to place various disease or impairment states along a common scale, should be considered further for measuring changes in the population's health status. Comparisons of competing approaches are just beginning. As research progresses, it may be possible to settle on a preferred approach. For the time being, however, experimentation with the various approaches just described should be encouraged.

Valuing Increments of Health

Economic analysis of health generally uses a compensating differential framework to estimate the values attached to better and worse health states. Ideally, situations are found in which people are trading health (or expectations about health) for income; monetary estimates associated with the tradeoffs are then used to infer the valuation of health. The voluminous economic literature on compensating differentials in labor markets provides a useful starting point. People make decisions involving risk comparisons all the time: what job to take and what safety devices to have in their car or house are two examples. In the classic case, the additional income one must pay to attract people to work in risky jobs is taken as a measure of the monetary cost associated with additional health risk. For example, to examine tradeoffs between current occupational hazard or risk and current wages, one might compare the wages paid to window washers who work at street level with the wages paid to window washers working on skyscrapers. The revealed willingness to pay in these settings is a measure of the value that people place on health risks. If the skyscraper job adds a 1-in-10,000 risk of death and pays $100 more per year than the street-level job that is otherwise similar in terms of skill requirements, intensity of the work pace, scheduling flexibility, working conditions, and so on, the implied value of the remaining life span of the person taking the riskier job is $1,000,000. Self selection makes the problem more complicated when aggregating across a population, as people taking riskier jobs will tend to be at the population margin in their valuation of life and/or level of risk aversion.

The multidimensionality of health dictates the way in which the compensating differential methodology can be exploited. There are so many health states that it is impossible, in practical terms, to find associated market conditions for each. Consider the situation of extending the life of retired people who have had heart attacks. There is no ready, comparable market situation we can imagine in which people face the risk of living after a heart attack. And even if there were approximations to the longevity benefits, the exact valuation would depend on such subtle factors as the severity of mobility impairment, interaction with other conditions that people have, and so on. The number of disease state and age combinations alone makes it clear that a contingent valuation cannot be established for every health state. This is why research has focused on valuing a common health state, a year in perfect health—the V in equation 6.1.

Viscusi and Aldy (2003) offer a comprehensive review of the literature that seeks to estimate the value of a statistical life. They find a great deal of variation in estimates of risk tolerance and, in turn, of the value of life. In examining more than 100 studies on mortality or injury risk premiums, the authors find the median value attached to a statistical life to be around $7 million for prime-aged workers in the United States. Reflecting the uncertainty of such estimates, however, the upper bound is higher than the lower bound by a factor of more than two at the 95 percent confidence interval (Viscusi and Aldy, 2003, p. 68).

While there is high variation in these kinds of estimates, and such values may not be applicable in all cases, they are probably reasonably accurate in many market-based settings. On a per-year basis, estimates in the literature translate into values in the $75,000 to $150,000 range (Kenkel, 1990; Murphy and Topel, 2003). In these kinds of studies, $100,000 for an additional healthy year is a typical value used.

> **Recommendation 6.8:** Recognizing that there is a range of uncertainty, a satellite health account should be based on a dollar figure for the value of a year in perfect health derived from estimates in the literature. That value should be updated as further research indicates.

Because of the controversial nature of valuation in this account, especially but not limited to valuation on the output side, measures quantifying physical changes in health status (e.g., quality-adjusted life-years) should be published alongside any monetary measures.

The high marginal value of life raises a fundamental question of affordability. Bounds on the value of life must ultimately be constrained by household budgets—willingness to pay must be anchored by ability to pay. Since, for most people, annual income is less than the $75,000 to $150,000 per person valuation figure used in the literature, people could not afford to buy many years of life before running out of income. How is this possible? The answer comes from recognizing the importance of initial constraints. The value of life as reflected in compensating variation studies is an estimate of the slope of the indifference curve at the current allocation. It answers the question of how much people would be willing to give up to get an additional year of life, given their current income and life expectancy. As the level of cash income and life expectancy change—for example, if one gives up more income to live longer—the indifference curve tangency point would change, and so would the value of life. With lower cash incomes and longer life expectancy, the marginal value of a year of life would fall. In this context, the value assigned to health should be allowed to increase over time along with income growth (see Costa and Kahn, 2003, for a discussion of rising life valuation).

Another set of issues arises in considering whether and how the value of life differs across people. These derived values of life clearly are related to income; rich people can afford to spend more on safety devices and medical treatment than poor people. One set of health accounts thus might value health to the wealthy more highly than health to the poor. Another might give equal weights. One can imagine circumstances in which weighing is appropriate, e.g., for measuring demand or for assessing the effect of health on productivity. In the latter case, one might want to weight by labor supply, or by labor income. Even in this case, if the expected incidence of disease were independent of socioeconomic status, the unweighted average value of health would be an appropriate measure to use. In other applications, such as for general measures of well-being or socioeconomic progress, equal weights are conceptually attractive. More generally, one can think about developing national health accounts from behind Rawls's veil of ignorance (without a priori knowledge of where they are going to be in the distribution of wealth or income, people will agree to an equal distribution). Suppose one were to value health knowing only what the prevalence of disease will be and the average income in society. In that case, the value of health would be the value to the person with median income. Alternatively, one may wish to focus on the health of the least well-off members of society or consider an index that focuses on those with the poorest health.

DATA REQUIREMENTS

Some of the data required for a satellite health account are already collected, but these data are not necessarily available in the form required for such an account. Before a credible health account can be produced, new data will be needed to measure the population's health status on a regular and continuing basis. For some years, the United States has had relatively good data on health status, such as those from the National Health and Nutrition Examination Survey (NHANES) and the National Health Interview Survey. Until 1999, NHANES—designed specifically to obtain information on the health status of the U.S. population—was conducted only once a decade. Now, however, it is an ongoing data collection program, and survey results are published more frequently; some are available annually. This is an important step toward developing a national health account. Output measurement will require more research that attaches meaningful physical health status data to monetary measures, including work to produce better data on the link between health and earnings (and vice versa). Data sources also will need to be developed that track changes in the relationship between prevalence of disease and years of healthy life and between medical interventions and health outcomes.

Health accounts also will require improved measures of the inputs to health. Better organized and more accurate data on medical care spending, aggregated by disease treatment, are part of what is needed. Improved data on care time, such as will be produced in the ATUS, also are necessary for developing the input side of the health account.

Measuring the quality of both the inputs to and the outputs of improved health is a further area for needed research. The statistical agencies are working to develop approaches for handling difficult-to-measure changes in the quality of health treatments. The Bureau of Labor Statistics, for example, is working on experimental medical care price indexes based on disease- and diagnosis-based units. Currently, data on medical care prices are organized primarily by institutional provider (e.g., payments to hospitals, doctors, or drug companies), not by treatment. If new treatments are developed for particular conditions that require fewer resources, this is not reflected in the form of a lower price level. Focusing on the cost of treating diseases or diagnoses allows prices to reflect changes in the mix of inputs used to treat particular conditions. Quality-corrected data on the full range of complete treatments do not yet exist—and as Triplett (2002) points out, we are still very far from having this information on per-case cost trends by disease. Such data would be extremely useful, even for conventional accounting of the medical care sector but also for the development of the health satellite account contemplated here.

7

The Government and Private Nonprofit Sectors

Measuring the value of public- and nonprofit-sector activities has long been recognized as a challenge. As with the other nonmarket areas considered in this volume, a common and fundamental difficulty is the absence of market prices, but there are also other conceptual issues that need to be resolved.

We examine the government and nonprofit sectors together because they are the loci for the production of outputs having sizable public goods elements, and it is these elements that pose a number of central measurement and valuation issues.[1] A key topic in this chapter is volunteer labor, which constitutes a large nonmarket input, especially in the nonprofit sector, but in government as well. This chapter does not address all of the ways in which measurement of governmental and nonprofit sector activity might be improved for specific types of applications. For example, we do not address in any detail the issue of how outputs that offset what would otherwise be negative effects on welfare should be accounted for—e.g., governmental resources going into added police services in response to an increase in crime.

Outputs of the government and nonprofit sectors commonly are given away, not sold. Examples include the provision of national defense and basic medical research by government, the research and dissemination activities of nonprofit universities, and the cultural and species preservation work of nonprofit museums and zoos. To finance such public goods (also sometimes termed collective goods),

---

[1]For an analysis of the roles of governmental organizations relative to private nonprofit organizations in public goods markets, see Weisbrod (1975).

governments typically levy taxes. The activities of nonprofit organizations are financed through some combination of private and government contributions of money, time (volunteer labor), and goods; tax subsidies; user fees; and assorted ancillary revenue-generating activities (e.g., university bookstores, museum shops, or hospital-operated commercial fitness centers).

In keeping with standard practice, the *outputs* of the government and non-profit sectors in the National Income and Product Accounts (NIPAs) are valued by summing the market value of the *inputs* purchased for their production (plus, for government, a capital depreciation component). Measured in this limited way—largely in terms of expenditures on labor and structures—output of federal, state, and local governments included in gross domestic product (GDP) was estimated to be $1.96 trillion of the total $10.49 trillion for national output in 2002 (Bureau of Economic Analysis, 2004). Over 22 million people are employed by government at all levels. This represents about 17 percent of all employment—a share that has held fairly steady over the past decade (U.S. Office of Management and Budget, 2003). According to Bureau of Economic Analysis (BEA) data, these government employees received more than $1.1 trillion in wages, salaries, and benefits in 2002 (Bureau of Economic Analysis, 2004).

Currently collected data also provide some sense of the importance of non-profit institutions in the economy. Most of what is known relates to organizations that serve the public and are classified under section 501(c) of the tax code. These organizations are eligible to receive tax-deductible donations and are large enough (over $25,000 in revenues) to be required to file annual reports with the Internal Revenue Service (IRS). Other organizations of particular interest, such as religious congregations and organizations with revenues of less than $25,000 per year, are not required to report to the IRS and, as a result, information about them is lacking (Boris, 1998).

Under law, there are many classes of tax-exempt "nonprofit" organizations. 501(c)(3) public charities include, among others, most nonprofits involved in the arts, education, health care, human services, and community service. 501(c)(3) private foundations are primarily grant-making organizations, such as the Ford Foundation or the Pew Charitable Trusts, that make grants to other nonprofit organizations. Other exempt organizations registered with the IRS include trade unions, business leagues, social and recreational clubs, and veterans associations classified under varying parts of section 501(c)(4) of the IRS code (Urban Institute and National Center for Charitable Statistics, 2000).[2]

Data published in the *New Nonprofit Almanac* (Independent Sector and the Urban Institute, 2001) provide an indication of the role that these organizations play in the economy. The Almanac focuses on groups it calls independent-sector

---

[2]For full definitions of these kinds of organizations, see the IRS web page on the topic, available at http://www.irs.gov/charities/index.html [accessed October 14, 2004].

organizations, which include those operating under sections 501(c)(3) and 501(c)(4) of the Internal Revenue Code, plus religious organizations. The total number of independent-sector organizations grew from 793,000 in 1982 to 1.2 million in 1998. In 1998, these organizations employed an estimated 10.9 million paid workers—about 7.1 percent of working Americans—and paid approximately $258 billion in wages (Independent Sector and the Urban Institute, 2001).

In the NIPAs, government activity is aggregated across industries, which serves to highlight its overall size as well as changes over time. While government purchases have long been aggregated to constitute a standard "final use" component of the national accounts, nonprofit activity is scattered across a number of industries. This has made it difficult to aggregate the data required to determine the size of and trends in the sector, even if the intent is simply to quantify market-based activity.

The BEA has historically maintained data on income and outlays of nonprofit institutions as part of the personal sector, which by convention includes nonprofit organizations serving households. Young (1993) provides a detailed description of how and where nonprofit organization activity is treated in the NIPAs and in the detailed input-output tables that underlie GDP. Recently, BEA has disaggregated the personal sector to provide separate estimates of expenditures, income, and savings for households and nonprofit institutions serving households. The new tables cover a wide range of institutions—religious, social services, medical care, education and research, recreation and some personal business—and these organizations' receipts and expenditures were reconciled with statistics on tax exempt organizations from the Internal Revenue Service. The BEA's preliminary estimates indicate that net current receipts of nonprofit institutions serving households—from sales of goods and services; transfer payments from businesses, governments, and households; and rental, dividend, and interest income—amounted to roughly $742.4 billion in 2001. This represents approximately 8.5 percent of total personal income, a share that has remained fairly constant since 1992 (Mead et al., 2003, p. 16). Nonprofits that serve business—credit unions, financial institutions, chambers of commerce, and trade associations, to name a few—are not included in the new tables (Mead et al., 2003, p. 13).

CONCEPTUAL FRAMEWORK

The government component of the national accounts has been improved in recent years. To take an important example, since their 1996 revision, the U.S. national accounts have distinguished between consumption and investment expenditures by governments. The accounts, however, still do not provide measures of the output of the government sector that are independent of the inputs, something that is essential to confronting a number of basic research questions related to the efficiency of government operations and other matters (National

Research Council, 1998, p. 2) What is the return on capital investment by the government? What is the return on government-supported research and development? Are government workers growing more or less productive over time? Do government enterprises—such as utilities, airports, transit services, schools, and hospitals—operate efficiently? How much aggregate savings occurs in the United States and how does government policy influence the savings rate? Meaningful answers to all of these questions will require independent measures of the output of the government sector.

Recent work to improve accounting for the nonprofit sector has been shaped by guidelines set down in the *Handbook on Nonprofit Institutions in the System of National Accounts* (United Nations, 2004). The stated goal of the U.N. project, and the presumed objective of a U.S. satellite account for the nonprofit sector, is to improve the ability to monitor the scope, size, structure, and financing of nonprofit activities and their effects on the economy. The U.N. work has focused primarily on measuring inputs—particularly volunteer labor—in a consistent manner across countries. The *Handbook* pays relatively little attention to the valuation of output produced by nonprofit organizations, discussing the matter but offering few recommendations.

It is easy to see why accounting for activity in the nonprofit sector is important and how it might be improved. The nonprofit sector is a large and growing component of many of the world's economies—accounting for perhaps 8-12 percent of paid nonagricultural employment in most developed countries and even larger shares in developing regions (United Nations and Johns Hopkins University, 2003, p. 3)—but the value it creates is not well represented in core national accounts. Additionally, provision of services by nonprofit organizations sometimes is viewed as a substitute for the services provided by governments, which means that government activity cannot be fully understood without also understanding activity in the nonprofit sector.

Nonprofit organizations share many common characteristics—nonprofit status, public-goods production, revenue structures that rely on donations of time and money, heavy use of volunteer staffing, and tax and legal treatment (United Nations and Johns Hopkins University, 2003). These characteristics argue for the analytic value of coherent data on the nonprofit sector as a whole, but both the System of National Accounts (SNA) and the NIPAs distribute the sector's expenditures and revenues across the government, corporate, and nonprofit institutions (serving households) categories (United Nations and Johns Hopkins University, 2003), making it difficult to obtain a unified picture of its activities.

Similar to their treatment of government, national accounting structures generally do not measure the output of nonprofit organizations independently from the inputs used in producing that output, something that should be a goal for a nonprofit satellite account. Furthermore, it may be beneficial to develop measurement concepts and classification structures for the nonprofit sector that are different from those used in the conventional NIPAs, and a satellite account would offer a vehicle for experimentation with alternatives. In both the government and the nonprofit sectors, the key challenges are how to define and measure "outputs" and then how to value them, recognizing that different objectives and uses imply different valuation approaches, so that there is no single "correct" approach.

> **Recommendation 7.1:** Measurement of government and nonprofit economic activity should be strengthened by developing satellite accounts that include and attempt to value nonmarketed inputs and outputs.

Because they would maintain both market and nonmarket components, these satellite accounts would not necessarily be fully compatible with the existing NIPAs. They would, however, expand the information base for studying changes in government and nonprofit activity over time and, possibly, making comparisons across regions and countries.

Experimental government and nonprofit sector accounting work should focus on developing new and quite possibly multiple, measures of output. For the nonprofit sector, the initial focus should be on developing an account for charitable organizations that are exempt from taxation—specifically those operating under section 501(c)(3), and perhaps some operating under 501(c)(4), of the Internal Revenue Code. These are essentially the organizations that are providers of public goods, rather than providers of private outputs to their members.

Nonprofit organizations that use mainly purchased inputs are similar to government entities in that the prices and quantities of these inputs register in the national accounts. Although not sufficient to answer all of the questions of interest, summing these up offers one measure of the organizations' contribution to GDP. For the satellite nonprofit sector account, a higher priority should be given to classes of organizations—particularly those in IRS subsection 501(c)(3)—whose activities are most seriously underrepresented in the national accounts because they involve large amounts of volunteer labor or rely heavily on in-kind donations.

The first step toward strengthening the measurement of nonprofit economic activity is to identify and organize the relevant market elements from the NIPAs, as the BEA has begun to do. These market elements then should be augmented with information on the nonmarket activities that are integral to the sector but are currently unmeasured or measured in ways that may be unsuitable for certain purposes. An important issue is how to value inputs that, like volunteer labor, are rationed by nonprice mechanisms, which in turn raises the issue of how to value the outputs produced by such inputs. Consumers generally are not free to acquire as much of a nonprofit- or government-provided good or service as they would like at the notional dollar price. This transaction price, which is often zero, thus may not accurately indicate the consumer's marginal valuation of the output.

Many of the inputs to the production of government and nonprofit organization production are purchased in markets: a university buys a computer, a church buys a new roof, or the Department of Defense buys new defense hardware.

These types of intermediate goods purchases and capital investments, already captured in the NIPAs, also should be included in any new satellite account that aggregates nonprofit sector activities. In addition, it is important to include in the satellite account information on government and nonprofit inputs that are not purchased in markets and, thus, not counted in the NIPAs. Examples include pharmaceuticals given by manufacturers to government and nonprofit health programs, groceries supplied for free or at reduced cost to soup kitchens, food provided by restaurants to volunteers who are donating their time, and computer hardware donated to schools. Most of these same goods are also sold in private markets and so, though they are unmeasured in the NIPAs, in principle they have observable prices. These prices may or may not provide a good indication of the value of items that have been donated. Consider, for example, donations of computer hardware or software to a school or donations of pharmaceuticals to a nonprofit health clinic. It would be questionable to value such donations at the prices at which the school or the clinic could have, but did not, buy the products. The observed market prices would probably overstate the recipient's willingness (and ability) to pay. Nonetheless, satellite accounts that include in-kind donations to government and nonprofit organizations could shed light on currently unmeasured economic inputs.

> **Recommendation 7.2:** Donations of labor and goods to government and nonprofit organizations should be characterized and described in quantitative terms, and approaches based on market comarisons should be developed for estimating their value as inputs.

Because some donated goods are first purchased in markets, the corresponding links with and effects on the NIPAs should be documented.

VOLUNTEER LABOR

Volunteer labor is the principal input to government- and nonprofit-sector activities that goes unmeasured in the NIPAs. There is considerable variation in measures of volunteer activity across surveys. Bureau of Labor Statistics (BLS) data from a supplement to the Current Population Survey (CPS) indicate that about 63.8 million people (age 16 and older) performed volunteer work from September 2002 to September 2003 in the United States. This translates into a volunteer rate of 28.8 percent among the civilian noninstitutionalized population. The median amount of time that people reported spending in volunteer activity for the period was 52 hours per year. For this work, volunteers are defined as "persons who did unpaid work (except for expenses) through or for an organization" (U.S. Bureau of Labor Statistics, 2003). Independent Sector and the Urban Institute (2004), based on data from their 2001 Giving and Volunteering Survey, report that 83.9 million Americans (age 18 and older) volunteered in 2000 (a volunteer rate of 44 percent), contributing an average of 3.6 hours a week, figures much larger than those based on the CPS data. Both of these estimates rely on answers to retrospective questions about activity over a 12-month period. The figures from the two sources have not been reconciled. When they become available, data on volunteer activity from the new American Time Use Survey (ATUS) should be more reliable than any that currently exist, but the ATUS data surely will not alter the conclusion that volunteer activity is significant in scope and magnitude.

According to the CPS data for the 2002-2003 period, the major types of organizations for which individuals volunteered were religious (34.6 percent), educational or youth-service related (27.4 percent), social or community service (11.8 percent), and hospitals or other health related (8.2 percent). The major activities performed included fundraising (28.8 percent), coaching, refereeing, tutoring, or teaching (28.6 percent), collecting, preparing, distributing, or serving food (24.9 percent), providing information such as by serving as an usher, greeter, or minister (22.0 percent), and general labor (21.8 percent).

Other more specialized examples of volunteer employment include accountants who help low-income people prepare their income tax returns and Earned Income Tax Credit applications, lawyers who provide pro bono legal services, and corporate executives who serve on the boards of directors of nonprofit organizations. In contrast to their counterparts on for-profit organization boards, even in the same industry, nonprofit board members often receive a negative wage in the sense that they are typically expected to give donations to the organization in return for the honor of board membership.

While survey data disclose important information on hours of volunteering and about the industries to which it is supplied, the meaning of the data is subject to interpretation. Even when asked specifically about volunteering in connection with an organization, for example, some respondents may perceive their participation in informal activities, such as a private quilting group or a poker club, as a volunteer activity. Examples of informal, nonmarket groups abound, some providing external benefits that may be of sizable consequence—e.g., "neighborhood watch" and community youth literacy groups—and others, such as local garden and investment clubs, providing benefits that are limited to members. These different cases suggest that multiple options exist for defining the "value" of organizations and for establishing the boundary of what is considered the output produced by volunteer labor. Furthermore, in some surveys, a great amount of what people report as "volunteering" may not be connected with an organization at all. In all of these cases, observed transaction prices of zero often mask a complex set of barter arrangements that yield explicit monetary prices of zero but that understate the private and social values being created (see below).

These definitional issues highlight the question of what should be covered in a nonprofit satellite account. Focusing attention in a satellite account on information about volunteering through or for an organization is defensible on practical grounds for initial forays into nonmarket accounting, both because of the interest that has been manifest in the activities of nonprofit organizations and because there is a reasonably clear boundary that can be drawn around this set of activities. At this time, the panel would prioritize valuation of volunteering to organizations, particularly those recognized as nonprofits by the IRS. As a long-term goal, a satellite account could be expanded to cover volunteer activity more broadly. To understand how the economy is functioning and changing over time requires measuring and valuing all productive unpriced labor time. Time spent helping others informally is arguably as or more important than formal volunteering. As the population ages, we might expect an increase in the reliance on friends, family, and neighbors to help out with grocery shopping, yard work, and other activities of daily life. Some of these activities fall conceptually between household production and volunteer work. The practical problems of measuring and valuing volunteer labor activity not connected with a formal organization may be severe; still, we want to highlight its omission from traditional economic accounts and, because such activity clearly contributes to real output, encourage attention to it.

We would also note, again, the potential for overlap. If a comprehensive volunteer labor account were developed, it would cover activities that also factor into other accounts proposed in this report. For example, time spent volunteering at a hospital is an input to health sector output, and time spent volunteering at a school is an input to education sector output. This potential overlap should not be avoided, but it means that analysts and users of data from the separate satellite accounts should be cautious to avoid double-counting of outputs.

Whatever accounting scheme is pursued, one must know the number of volunteers and the number of hours they worked in order to derive an estimate of total volunteer hours. One also needs to know the nature of volunteer activity and the values to be assigned to the hours of each type of work performed. It is reasonable to expect that good estimates of the number of hours worked by volunteers can be obtained—as already noted, the ATUS should be enormously helpful in this regard.

The appropriate valuation to be attached to an hour of volunteer labor depends on the purpose of the measure being constructed. An hour of volunteer labor implicitly enters the national income accounts with a price of zero, just as other inputs enter at the prices paid for them. If the goal is to measure productivity or economic welfare, however, attributing a value of zero to volunteer labor has little appeal. If, for example, the supply of volunteers increases in response to shrinking opportunities in the paid labor market, a zero valuation of volunteer time would bias measures of change in economic output. The NIPAs would reflect the decrease in paid employment, but they would not capture any increase in production associated with the increase in unpaid volunteer work. Thus, a shift of workers between paid and nonpaid employment would, other things equal, alter measured national income and product in a manner that distorts the true change in economic welfare and economic growth. It would exaggerate the *decline* in economic welfare if the economy is contracting and labor shifts from the paid labor market to the volunteer market, and it would exaggerate the *gain* in economic welfare if the economy expands and there is a shift in the opposite direction. This is very much like the well-recognized distortion in national income trends accompanying shifts from household production, such as home-prepared meals, to market production, such as restaurant meals, or vice versa.

Valuing the contributions of volunteer labor to government or nonprofit activity at zero, as in the national accounts, masks a great deal of complexity. The *total* contribution of this volunteer labor to economic output, and to welfare, clearly is not zero, and, indeed, even its *marginal* value very likely is positive. If a cost-minimizing organization—government, nonprofit, or for profit—can obtain as much of an input as it likes at a zero price, it can be expected to increase utilization of the input until its marginal productivity falls to zero. But the supply of volunteer labor is not unlimited—organizations typically cannot obtain as much of it as they wish. Thus, the observed market price of zero likely understates the marginal productivity of volunteer hours. In this case, valuing volunteer time at its observed price produces downward-biased estimates of its contribution to economic output.

If one is interested in the full welfare contribution of volunteer work, a different approach that accounts not only for the value of volunteer services to the receiving organization but also for the utility that volunteers derive from the activity may be appropriate. Indeed, in addition to providing experiential benefits, some volunteer activities (e.g., pro bono legal work) may enhance a person's professional status and possibly raise his or her future earnings. People who obtain utility from volunteering will be willing to work at a monetary wage of zero despite having a positive opportunity cost of time. This is not so different from the market case in which an individual, confronted with two job choices, accepts the lower paying option because it is expected to provide more nonmonetary benefits.

Interestingly, paying a wage of zero gives the "employer" an implicit exemption from minimum wage requirements. If an organization were to seek to increase the amount of labor available to it through volunteer channels by offering an explicit wage greater than zero, it would face a significant discontinuity, as it would be required to jump from paying a zero wage to paying at least the minimum wage.

There are at least two possible goals that an analyst might have in mind when calculating the value of volunteer labor. The goal that is chosen will affect the approach adopted. One goal might be to measure only the contribution of the labor input toward the production of organizational goods and services, however those are valued. Alternatively, the goal could be to capture the full welfare effect of volunteer activity, including any utility flowing to the volunteer in addition to the benefit to the organizations to which time has been donated. The latter is certainly a more ambitious goal, and one that introduces the need for additional assumptions and information.

Adopting an approach to the valuation of volunteer labor that is consistent with the approach we have recommended for the household production account (see Chapter 3) would imply that one is interested only in the contribution of volunteer labor to the production of goods and services by the organization to which the volunteer labor is supplied. In this case, it seems natural to value volunteer labor at the cost of hiring someone to perform the same tasks, adjusting for possible productivity differentials between paid and volunteer labor, rather than at the opportunity cost of the person providing the volunteer services.

Applying the replacement cost method for valuing volunteer labor requires the identification of employment categories from which paid replacements could be drawn. In the past, this commonly has been done using very blunt wage information—e.g., the average hourly wage rate in the economy. A more targeted and useful approach would use the wage earned by workers who are hired to supply a similar kind of labor (e.g., using the wage of paid teacher aides to value the time of classroom volunteers). For volunteers who donate services that they also sell in commercial markets—such as accountants and lawyers who provide pro bono hours—these individuals' market wage rates might be a reasonable estimate of the relevant replacement cost.

For the household production account, the panel advocates a modified replacement cost approach (see Chapter 3), and there is considerable appeal to using the same approach for the almost parallel situation of volunteer activity. The idea of adjusting the wage rate to reflect systematic differences in the skill and productivity of paid versus volunteer labor is conceptually akin to the idea of adjusting market wage rates to account for the relative productivity of home and market producers. Volunteers may not have the same level of specialized skills as paid workers; the work environment for volunteers may be quite different (e.g., less formal) from that of paid workers, even within the same organization; and, perhaps most importantly, we do not know the extent to which volunteers and paid workers with similar job titles actually perform the same tasks. For example, do volunteer fundraisers, firemen, teacher aides, and hospital "candy-stripers" have the same levels of responsibility, skill, and productivity as their paid counterparts? If there are reasons to believe that volunteers are generally less productive than their paid counterparts, the wages of paid labor in a particular job category would be an upper bound on the market value of volunteer time. The legal minimum wage would represent a lower bound on organizations' cost to replace volunteers with paid labor, although there is no assurance that a user of volunteer labor would be willing to pay even the minimum wage for the quantity of volunteer labor being used.

It is worth noting that the *Handbook on Nonprofit Institutions in the System of National Accounts* supports an approach to valuing volunteer labor that is similar to this panel's recommendations relating to unpaid work in the household (United Nations, 2003, p. 73):

> The recommended procedure for the NPI satellite account is . . . a form of the replacement cost approach that ideally uses as the shadow wage for volunteers the average gross wage for the occupational activities in which the volunteers are involved, taking account of known large discrepancies in the skill levels of paid employees and volunteers. Since this requires more detail on the activities in which volunteers engage than is likely to be available in most countries, however, we recommend a fall-back approach that assigns to volunteer hours the average gross wage for the community, welfare, and social service occupation category.

The SNA choice of the fall-back occupation is rightly conservative—the wage rate for the "community, welfare, and social service" occupation category is below the average wage rate in most countries. But using this approach as a basis for valuing volunteer labor, while operationally appealing, has little conceptual justification. Moreover, it assumes that volunteers represent a cross-section of types of labor utilized in paid jobs, a matter about which little is known.

The welfare-based valuation of volunteer time can be examined in the framework of a model in which a person seeks to maximize utility (U) by allocating time among three uses: leisure, paid work, and volunteering:

$$U = U(L, E, V) \tag{7.1}$$

where L is hours of leisure, E is hours in paid employment, and V is hours of volunteer work. Time uses L and V are positively related with utility ($\partial U/\partial L > 0$ and $\partial U/\partial V > 0$); hours in paid employment typically are assumed to have a negative relationship with utility ($\partial U/\partial E < 0$), though some people certainly may enjoy their market work effort. The three uses of time must all be non-negative and must sum to total discretionary time available. In equilibrium, a person seeks to equate the marginal value of time in all three uses, subject to the non-negativity constraint. The marginal value of time devoted to leisure or volunteer work is just the marginal utility associated with the activity; the value of market work equals the wage that is earned by working less the disutility, if any, associated with the performance of that work. A person may choose to perform no paid labor, which is typical of retired people, or to engage in no volunteer work, which characterizes about 70 percent of all adults.

For persons who work and also volunteer, which describes some 30 percent of paid workers (U.S. Bureau of Labor Statistics, 2003), the equilibrium values of an hour of both V and L should be equal—to each other and (in the absence of marginal utility or disutility from time spent in market work) to the marginal wage, W, in the paid labor market. W is observable and provides a basis for imputing a value, at the margin, to volunteer time.

In this framework, volunteering contributes to economic welfare in two ways—as a labor input to production and as a provider of utility to the volunteer.

A full accounting for the contribution of volunteer labor to economic welfare would identify, measure, and sum these values. In the simplest case, the value to the individual of participating in volunteer activity is just the foregone wage. If volunteers actually receive a negative wage by incurring costs (of travel, meals, uniforms, requirements to donate, and so on) in return for the opportunity to engage in volunteer activity, then the volunteer is deriving utility that exceeds the market wage that is foregone.[3] Viewed in this way, the total contribution of an hour of volunteer labor to economic welfare is the sum of the positive value accruing to the recipient organization and the potentially even larger positive value accruing to the volunteer.

Conceptually, there is merit to the method that reflects the value the volunteer places on the opportunity to participate in volunteer activity. Examining the activity as a barter arrangement, in which the organization derives a marginal product of labor and volunteers derive utility, may reveal information about the value of the volunteer time to the employing organization as well as to the volunteer. At this time, however, the methodology to produce such estimates is not developed, and, while this direction for future research should be explored, it is difficult to envision how reliable data for this approach would be produced.

> **Recommendation 7.3:** Volunteer labor should be valued, at least initially, by finding the appropriate market analogues—wages paid to employees working in similar occupations. To the extent possible, this replacement cost approach should be modified to reflect skill and effort differences between volunteer and paid labor.

If productivity adjustments cannot be made, we suspect that the replacement labor wage generally will represent an upper bound on the value of volunteer time inputs. This is a subject on which additional information certainly would be welcome.

Recognizing that there are many uses of this kind of data and that no one approach is best for all of them, we take the position that multiple approaches should ultimately be accommodated. For example, given that both a replacement cost and an opportunity cost approach may be useful, the satellite account could contain two or more sets of value data. Additionally, the panel favors the publication of information that presents quantities (person-hours) of volunteering, by type of work performed and by type of provider organization (specific industry, government, or nonprofit), separate from the values, so that each may be used as needed for a particular application.

---

[3]Some of these costs of volunteering are deductible on individual income tax returns; nevertheless, the volunteer bears most of the costs.

DONATED GOODS

Labor is not the only input that is donated to nonprofit or public organizations and that, as a result, is unpriced and unmeasured in the NIPAs. Many of the issues discussed above in the context of volunteer labor also apply to donations of goods. Estimating replacement costs to recipient organizations is an option for valuing donated goods, just as it is for valuing volunteer labor. It should be noted, however, that just as in the discussion of how to value volunteer time, it is by no means clear that recipient organizations would, in the absence of the donations, purchase the donated goods at anything close to the observed market prices. Thus, market prices are likely to be upper bounds on the values to recipient organizations of donated goods. In contrast with the volunteer labor analysis, however, it is less obvious that corporate donors, or perhaps even individual donors, derive utility from the act of donating goods. For corporate donors, such donations could represent investments in goodwill and future sales, and the return on these investments would eventually appear on the donating companies' books as market transactions.

Scant attention has been given to the magnitudes or values of goods donated to nonprofit and public organizations. The values of donated goods as reported on itemized individual tax returns seem likely to be overestimates of market values. For example, a recent report by the U.S. General Accounting Office (2003) looked at donations of automobiles to charities and concluded that the amounts deducted on itemized tax returns by donors substantially exceeded the amount of money realized by the charities when the cars were later sold. The values of corporate donations of goods also may well be upward biased insofar as they reflect corporate tax laws that permit deductions based either on market prices for the donated products or on average production costs even when marginal costs are far lower. Thus, the market price of goods donated to public or nonprofit organizations is an overestimate of what would be purchased at that price, and allowable tax deductions for donated goods are also overestimates of the marginal cost of producing the donated goods. Still, if these measures of the value of donated goods could be obtained from the IRS or other sources, they would constitute at least a first step toward extending the coverage of satellite accounts to include goods donated to nonprofit and public organizations. More refined measures can and should be pursued over time.

MEASURING AND VALUING OUTPUT

What are the outputs of the government and nonprofit sectors? These sectors are dominated by services—and difficult ones to measure at that—education, health, social services, culture and the arts, recreation, and others. As noted elsewhere in this report, a major challenge in measuring the output of such services is simply to define the unit of output that is being transacted (Griliches, 1992).

Government spending represents about 18 percent of U.S. GDP. These expenditures go toward provision of goods and services ranging from defense to the judicial system to education. GDP excludes the transfer payment components of government spending (social security, welfare benefits, unemployment benefits, and so on).

Statistical agencies have made some attempts to measure government output directly rather than indirectly. Jenkinson (Organisation for Economic Co-operation and Development, 2003) offers a summary of international efforts, most of which aspire to the SNA recommendation to pursue direct volume measures, as opposed to the traditional input-based methodology. Atkinson (2004) provides an interesting review of the United Kingdom's efforts in this regard. Sectors typically covered as countries begin to develop direct measures of government output include education—where measures such as number of pupils and pupil hours are calculated—and health—where indicators include such things as number of patients and hospital treatments. But valuation of these outputs is difficult. Observed prices, even when they are positive, as with hospital charges to health care insurers, may well be of limited usefulness. In the case of health care, for example, because patients confront far lower marginal prices than do insurers but also confront a variety of nonprice rationing mechanisms, the marginal value to the patient of treatment provided could be either lower or higher than its cost. In work discontinued in the mid-1990s, The BLS developed measures of output for selected federal government agencies as part of its government productivity measurement program. These output measures typically relied on quantity indicators of one sort or another (Fisk and Forte, 1997). Work to measure government output for the United States is currently under way at the BEA (see Fraumeni et al., 2004); the goal of this work is monetary valuations.

In a similar spirit, the SNA categorizes physical output measures for 11 nonprofit sector groups (United Nations, 2003):

- culture and recreation;
- education and research;
- health;
- social services;
- environment;
- development and housing;
- law, advocacy, and politics;
- philanthropic intermediaries and volunteerism promotion;
- international;
- religion; and
- business and professional associations, unions.

Each of these groups contains numerous fields, subfields, and target physical output measures. Several countries—including Australia, Belgium, Canada, Italy,

Mozambique, the Netherlands, the Philippines, South Africa, and Sweden—have begun to develop new accounts intended to estimate the size of the nonprofit sector and to support international comparisons of nonprofit activity. Much of this work is progressing within the framework provided by the *Handbook on Nonprofit Institutions in the System of National Accounts* guidelines.

In some situations, outputs of governments or nonprofit organizations are not sold at a positive price because they are difficult or impossible to sell—as is the case with collective goods such as national defense, basic medical research, and environmental protection. In other cases, when outputs could be sold (e.g., education), a social judgment has been made not to do so. In still other cases, outputs are sold, but at prices not covering their full costs—e.g., service on some mass-transit systems and, in the nonprofit sector, college education even for the typical student but especially for those with scholarships. There is nothing necessarily inefficient about such pricing policies, but the social valuation of the associated outputs remains a vexing problem.

When the output is a purely collective good, it is by definition infeasible or inefficient to exclude particular individuals from its consumption. There is no practical way, for example, to exclude any person from the benefits of national defense, species preservation, or locally, clean air. In some cases, such exclusion may be technologically feasible, as with public radio or public television. The Public Broadcasting System could institute pay-per-view charges for particular programs, as sometimes is done by commercial vendors. Even when feasible, however, such pricing may be economically inefficient because there is no marginal cost of allowing an additional person to have access to the service.

The Zero Price Problem

How should one think about a price of zero? A price of zero could imply a marginal value of zero, as is surely the case for a vast number of potential outputs that are not provided at all because there is little or no demand for them. But a price of zero for pure public goods is another matter. Though observed prices are, and to be efficient should be, equal to zero, to conclude that such outputs have no value to society does little to inform research in most policy areas. Likewise, the often-invoked convention of valuing output on the basis of prices paid for inputs, whether positive or zero, is largely arbitrary. Under conditions characterized by positive "congestion costs"—when consumption by one person imposes external costs on others (as on crowded roads and in air-polluted areas)—efficient prices would be greater than zero. The absence of estimates of these values is a current shortfall of the NIPAs, and more attention to them and to their changes over time and across areas would be useful.

The key issue here is the relationship between the valuation of outputs at their market prices and the economic welfare interpretation of those values. Market prices (and the associated quantities) have general utility as an indication of the value that purchasers attach to the marginal output. When explicit prices are not charged for output, as is common in the government and nonprofit sectors, buyers' valuations of the outputs are not available.

It should be noted that, even when market prices are available, they reflect buyers' willingness to pay for a marginal unit of output, not the willingness to pay for the entire value to them of the good or service. The difference between the two values, a matter discussed in earlier chapters, is attributable to the inclusion of consumer surplus in the latter but not in the former. For aggregate welfare measurement purposes, and to measure changes in welfare over time and across countries, information on consumer surplus is conceptually both relevant and important; the inclusion of consumer surplus in a satellite account for government or the nonprofit sector, however, would make that account incompatible with the NIPAs.

Even with a narrower focus on marginal valuations, problems associated with outputs having prices of zero remain. One alternative is to value such outputs by using the prices paid for the inputs to their production. When outputs are not sold, it is nonetheless true that their production typically requires purchased inputs of labor, raw materials, and capital (in addition to such donated inputs as volunteer labor). Aggregating market input values is, in fact, the methodology used to value the output of most government as well as nonprofit-sector services but, as we explain above, this procedure can be thought of as a pragmatic compromise when the alternatives are either to value those outputs at zero because they are not sold in the market or to value them at some positive, imputed, amount.

Asserting that outputs have values equal to the market values of the inputs with which they are produced imposes a strong, controversial assumption about output beneficiaries' willingness (and ability) to pay. Whether incremental expenditures on national defense, basic research, charity services to the poor, or roads are "worth" their cost—not more, not less—remains a subject of active debate.

Ruggles (1983) defends the idea of measuring government output by the market value of inputs. The claim is that, in a democratic society, government will supply goods and services in amounts that reflect citizens' willingness to pay taxes for their provision. It would follow that the total value of output is at least equal to the cost of the inputs. This perspective deserves further attention, but it is by no means unproblematic. It assumes that voters are well informed about the values of public-sector outputs, have the choice to vote on specific "earmarks," and do not engage in strategic behavior, depending on the specific tax mechanism expected to be used to finance the activity. It also assumes consumers' knowledge of that finance mechanism. In addition, this perspective relies on consumers' (voters') assessments of the total value of the service output, not the marginal values as would be captured by private market prices.

Whatever the merit of Ruggles's political-economic rationale for valuing government output by the cost of the inputs used to produce it, essentially the same arguments and limitations apply to the valuation of nonprofit-sector activity. Instead of relying on majoritarian voting processes, for the nonprofit sector, one would rely on donors' willingness to contribute and, hence, to pay for inputs purchased in ordinary markets, as indicative of the value of outputs to them. If this approach were adopted, one would also need to account somehow for output financed by profits generated through user fees and auxiliary activities, which vary greatly across industries. Also, "free-rider" behavior, leading to suboptimal donations, can be expected to generate expenditures on inputs that fall short of efficient levels, although tax subsidies to donee organizations (exemptions from taxes on property, sales, and profits) and the deductibility of donations on personal income tax returns exert a countervailing effect. It is likely that, in the nonprofit sector, the observed levels of output are short of those at which the aggregate willingness to pay for marginal output equals the marginal supply cost, but this is not certain.

What, then, are the options for measuring the value of nonmarketed outputs or, more generally, outputs that, if not provided at a price of zero, are offered at subsidized prices? Even when market prices are available as indicators of value, the appropriate measure depends—as we note throughout this report—on the use to which an output measure is to be put. Information needed for benefit-cost analyses is not the same as that needed for assessing growth of output or of economic welfare over time. Both differ from the data captured in national accounts that are based primarily on transactions prices and that in many cases value inputs and outputs at zero when no explicit prices are paid. To see this, consider cases in which public policy involves decisions about the efficiency of a governmental action—a plan to construct a dam, a new environmental pollution regulation, or an increase in funding for basic science research through the National Institutes of Health or the National Science Foundation. The benefit component of benefit-cost analysis requires, conceptually, valuation of the aggregate willingness to pay for the proposed outputs, amounts that encompass the total area under all beneficiaries' demand functions, not simply estimates of the marginal value to society of the dam, the environmental protection, or the research.

Acknowledging the difference between measurement for benefit-cost analysis and measurement for consistency with national economic accounting is critical for both the government and nonprofit sectors. Benefit-cost analysis addresses an explicitly normative question of whether a particular output is "worth" the cost of producing it, which requires measuring the total value of a specified quantity of a public- or nonprofit-sector activity for comparison with the associated cost. To this end, the value of public goods can be estimated by the sum of consumers' willingness to pay as derived from surveys that would in principle, but with great complexity in practice, capture consumer surplus, or by some imputation from observed behavior.[4] By contrast, measurement of the output of public goods that

[4]The survey approach to output valuation for goods and services not generally provided in private markets has been used to estimate values of cleaner air and community ambulance services, but it remains a controversial methodology. For an overview of the debate on contingent valuation, see Portney (1994) and the other articles in the same issue of the *Journal of Economic Perspectives*.

is consistent with NIPA concepts requires imputation of marginal willingness to pay for the quantities being supplied.

All of these options for valuing government and nonprofit-sector nonmarket outputs are problematic. Nonetheless, there is no justification, except on grounds of convenience and consistency with the use of observed market prices, for placing values of zero on nonmarket activities. If this approach were followed, the experimental satellite account would provide no new information relative to the NIPAs, and there would be little point to pursuing its construction.

Value Imputations

In the government and nonprofit sectors some outputs are sold, while others are not. In the hospital and nursing home industries, for example, as well as in private higher education, outputs are sold to some consumers, and at multiple prices, but given away to others. In such cases it is standard practice under generally accepted accounting principles (GAAP) for nonprofit organizations, in their filing of IRS Form 990 returns, to "gross up" the charity services of health care or education that they have provided. That is, organizations report not the realized revenue but the revenue that would have been realized had the free care or education instead been sold. This accounting practice, involving the imputation of revenue to outputs not actually generating revenue, assumes implicitly that the value of the charity care (or low-cost education) to the consumers who do not pay for the services they receive should be estimated by the value of similar services provided to the consumers who do pay. It is clear, of course, that the charity patients (or scholarship students) are typically not, in fact, willing to pay—the willingness to pay concept encompasses the ability to pay as well—the market prices. In that sense the imputed values overstate market values. The GAAP convention on imputation of charitable health care and education activities has been developed with respect to private nonprofit organizations, yet the fundamental issue it addresses is far broader. The grossing up of revenue, or more generally of output, to value unpriced output poses conceptual problems of value imputation at both the private accounting and national accounting levels.

Imputation is not a process that is used consistently in the NIPAs, but neither is it unprecedented. As is noted throughout this volume, some values are imputed—for example, unobserved values (annual flows) of owner-occupied housing services. The potential for making greater use of imputations remains. For nonprofit-sector accounts, the usefulness of imputing values to outputs exchanged at prices of zero, and the alternative mechanisms for making the imputations, deserve more attention. For the government accounts, the imputations for services of capital, which now are assumed to equal depreciation, might in the future include a net return (which separates out depreciation). For governmental and nonprofit organizations, the research knowledge base is currently insufficient for identifying a best approach for imputing values to unpriced outputs—in part because there are multiple uses for such imputations. Priority should be given to research that would lead to professional consensus on a small number of approaches for imputing values to unpriced governmental and nonprofit-sector outputs, and to developing methodologies for implementing each.

DATA REQUIREMENTS

The BEA is already using a broad array of data sources to account for income and outlays of nonprofit institutions that serve households. Data on expenditures of nonprofit institutions originate from the Census Bureau's quinquennial economic census and the annual economic surveys. Additional data sources are used to fill gaps for labor, political, religious, and educational organizations. The Urban Institute's National Center for Charitable Statistics disseminates information about the reports that tax-exempt institutions file with the IRS concerning their income, expenditures, and activities. Other data come from the American Association of Fundraising Counsel's Trust for Philanthropy and surveys of charitable contributions to religious and other nonprofit organizations conducted by the Independent Sector.

To push research further, we note five areas in which further data development activity would be helpful. First, for documenting activity in the nonprofit sector, improved access is needed to existing administrative records—including forms that organizations file with the IRS, as well as other financial statements and records. The Urban Institute's recent initiative has made the IRS Form 990 information easily accessible for anyone with an Internet connection, but there is other useful information that remains inaccessible. As an example, we support permitting access by researchers, under appropriate safeguards, to the confidential Form 990T data on commercial activity by nonprofit organizations that is "not substantially related" to its tax-exempt mission.[5]

Second, coordination of and access to data maintained by various nonprofit and government organizations can be improved. A good example of a situation that should be addressed is the lack of coordination between the BLS and the Census Bureau business lists. The Confidential Information Protection and Statistical Efficiency Act of 2002 (CIPSEA) has facilitated some information sharing among the statistical agencies for statistical purposes. Unfortunately, until companion legislation that modifies existing IRS statutes is passed to allow the Census Bureau to share information based on tax records with other statistical agencies, much useful work will be stalled.

---

[5]The Form 990T is required for all nonprofits that have revenue from "unrelated business income." The Form 990T is, in effect, a profits-tax return. The Form 990 includes reported gross revenue from unrelated business income activity, but not costs or net revenue (profit). There has been considerable use of Form 990 data for research purposes, and some use of the confidential 990T data under special arrangement with the IRS.

Third, improvement is needed in the coverage of giving and volunteering in household surveys. The private surveys that have been conducted generally have small sample sizes and relatively low response rates, and they are not conducted on a regular basis. Even the CPS supplement data cited earlier may be subject to bias related to the length of the recall period for which respondents are asked to report their volunteer activities. The ATUS should be a boon to researchers wanting to compile data on time spent in volunteer activities.

Fourth, new surveys of nonprofit institutions could provide much valuable information. Two types of additional information would be particularly useful. On the input side, existing data on volunteering have been derived almost exclusively from household surveys. It would be valuable to have data on volunteering from the perspective of the recipient nonprofit and government organizations. How much volunteer time is used by organizations of various sizes and in various industries? Would the organizations use more volunteer time if it were available—that is, are they supply constrained? What would it cost to hire people to perform the work now done by volunteers? Would they hire the replacements if the volunteers were not available? What is the most they would pay if they had to hire them? Turning from volunteers to other unpriced inputs, what kinds, and how much, of other donated inputs do they receive—food, equipment, office or production space, and so on? On the output side, how do organizations measure their performance? How do they, or would they, estimate the value of the outputs? Whether the typical nonprofit organization would be able to answer all of these questions is uncertain, but there would be value in attempting to learn what information they can provide.

Fifth, it would be useful to identify and develop data that create options for valuing hours spent in volunteer activities. Shadow wages based on similar paid occupations would be a sensible starting point, but research to assess the relative productivity of paid and volunteer labor performing similar tasks also would be worthwhile.

CONCLUSIONS

In considering how to progress with research to develop government and nonprofit satellite accounts, it is important to keep clear the distinction between what is optimal conceptually versus what is the best that can be done operationally. This chapter has only touched on the complications inherent in the construction of these satellite accounts. Pieces of the puzzle are within grasp. This report, and others, have described how a nation's volunteer labor inputs could be counted and valued. Such data would be useful to research in a number of policy areas—for example, that aimed at improving provision of public services and defining the role of the state—even if it is missing the output side valuation needed for a fully specified income and product account. It is realistic to believe that an account can be organized to provide a more comprehensive picture of the market inputs and outputs associated with legally defined private nonprofit organizations. As noted earlier, BEA is already working to produce new aggregate tables of nonprofit-sector activity. Such data will improve policy makers' ability to track the role of nonprofit institutions in providing services in specific areas such as education and health.

At this point, however, full and independent (non-input based) valuation of goods and services produced by government and by the economy's nonprofit institutions remains a long way off. For the foreseeable future, national accountants and other data producers will have to choose from more standard options: valuing output at exchange prices (often zero); valuing output based on the cost of inputs; or imputing values based on GAAP (the generally accepted accounting principals), which attempts to approximate the prices at which the outputs of nonprofit organizations in the education and health sectors could have been sold. In some cases, practitioners may choose simply to provide physical indicators of output.

One obstacle to producing a double-entry accounting of the nation's nonprofit activities is that they defy easy categorization. Nonprofit organizations and volunteer labor operate conspicuously and independently across many sectors of the economy. Volunteer labor is offered not just to nonprofit institutions, but also to government and other organizations, and nonprofit as well as governmental organizations utilize both paid and volunteer workers. All of this makes it problematic to prescribe a unified approach.

Much of the activity of the public and nonprofit sectors involves preservation and development of human capital. Conceptually, the output associated with public and nonprofit inputs to health should be valued in ways consistent with the methodology prescribed for the health account. A different methodology would be used for education and each of the other areas in which government or nonprofit institutions are active. In a real sense, development of the experimental health and education accounts is a prerequisite to the development of more comprehensive government or nonprofit satellite accounts.

The principal contribution of a satellite government account is the opportunity to make a bolder and more experimental stab at independent valuation of government output. As just indicated, work to construct such an account necessarily will draw from counting and valuation methodologies used in the accounts for the various sectors in which there is a strong government presence. As output valuation methods improve in health, education, and other services, the ability to estimate government output independently by definition also will improve.

A nonprofit-sector satellite account would require, at a minimum, a selective rearrangement of data for education, health, and other social services, and would entail major overlap not only with satellite accounts for activities in these areas but with market accounts as well. As with the government satellite account, a nonprofit account could embody numerous valuation methodologies. This fact underscores the panel's view that there is no single right way to present input and output data. In principle, data could be presented at an "atomic" level, such that the atoms could be aggregated in any way the user wished. As a practical matter, decisions must be made about which aggregations are worth making, as has been done historically for the NIPAs and will need to be done for the satellite accounts. Categorization by industry has proven useful, as has the consumption and investment delineation and the separate presentation of information for government and private entities. In advocating the development of a nonprofit satellite account, we are in essence making the case for expansion of the ownership criterion to include separate designation of the nonprofit form. For some purposes, one may wish to know, for example, the size and activity levels of the hospital and education industries, and their changes over time. For other purposes one may wish to know the size and activity level of the government, nonprofit, and for-profit sectors across industries. Viewed in this way, detailed information about industries does not make the government and nonprofit accounts redundant with other nonmarket accounts, such as health and education.

8

The Environment

Among the earliest, and now most thoroughly investigated, applications of nonmarket accounting were efforts to monitor change in the natural environment. While neglect of the natural environment in economic measurement concerned economists even in the nineteenth century, serious work to modify the conventional accounts did not begin until the late 1960s. Since then, there have been many official (governmental) and unofficial attempts at environmental accounting—those by the United Nations and the Bureau of Economic Analysis (BEA) have been among the most visible. The United Nations (1993) produced a *Handbook of National Accounting: Integrated Environmental and Economic Accounting* (referred to as SEEA).[1] In the United States, BEA began intensive work on integrated environmental and economic satellite accounts (IEESA) in 1992, but in 1994 Congress directed BEA to suspend that work (see Bureau of Economic Analysis, 1994a). That such efforts have been undertaken reflects the view that there are benefits to be gained from developing natural resource and environmental accounts.

A previous report, *Nature's Numbers,* which describes and critiques the major environmental accounting efforts to date, identifies these benefits (National Research Council, 1999, p. 2):

> [Environmental and natural resource accounts] provide valuable information on the interaction between the environment and the economy; help in determining

[1]An updated version (2003b), which has not undergone final editing and reproduction, is available at: http://unstats.un.org/unsd/environment/seea2003.pdf.

> whether the nation is using its stocks of natural resources and environmental assets in a sustainable manner; and provide information on the implications of different regulations, taxes, and consumption patterns. . . . More generally, augmented NIPA that encompass market and non-market environmental assets and production activities would be an important component of the U.S. statistical system, providing useful data on resource trends.

This panel agrees with the basic messages of *Nature's Numbers*, particularly the more overarching recommendations, which we strongly endorse.

> **Recommendation 8.1:** A concerted federal effort should be made to identify and collect the data needed to measure changes in the quantity and quality of natural-resource and environmental assets and associated nonmarket service flows. Greater emphasis should be placed on measuring effects as directly as possible, particularly on measuring actual human exposures to air and water pollutants (National Research Council, 1999, pp. 7-8).

The obvious place for new environmental accounting work to recommence is within the BEA, though the effort might equally well be housed at the Environmental Protection Agency, conducted with expert guidance from the BEA. Given this panel's general agreement with recommendations published in *Nature's Numbers*, the discussion of environmental accounting in this chapter is brief, focusing on the analysis, conclusions, and recommendations from that study.

DEFINITION AND SCOPE OF COVERAGE

Environmental accounting generally focuses on the measurement of natural or environmental wealth and the goods and services generated by this wealth. Accounting efforts have encompassed both current accounting and capital accounting—current accounting of the generation of pollutants or of the consumption of natural resources and capital accounting of changes in the stock of natural resources or the condition of the natural environment. For the most part, natural and environmental wealth excludes assets that are man made. Some environmental accounting systems—in particular, the French patrimony accounts—also include certain structures of special historical or cultural importance, such as ancient monuments. Examples of environmental goods generated by natural wealth are timber products and minerals. Examples of environmental services include recreational services, life support of animal and plant species, and esthetic services. One of the most important services is waste disposal, a positive benefit that is often accompanied by a "bad" or "disservice," conventionally referred to as pollution.

Valuing nonmarket natural assets is not an easy matter. Some environment-related goods, such as forestry products or mined minerals, are sold in markets and therefore pose no major problems; but others, such as clean air and water or unproven reserves of natural resources, are not priced in markets. A full accounting of natural resources and the environment must recognize that these assets, even those with market production costs of zero, have value, and that their value can be affected by human activities (Nordhaus, 2004, p. 12).

Market and Nonmarket Factors

At the outset, it is important to note that environmental and natural resource accounting is not necessarily the same as *nonmarket* environmental and natural resource accounting. Much environmental wealth, and the goods and services this wealth generates, is already measured in conventional economic accounts. Measurement is quite likely when legal, institutional, or technical arrangements permit the marketing of assets and any generated goods and services. Such marketing can be expected when the assets or the goods and services are privately owned. But marketing can take place even when the asset is in the public sector. A forest may be a public asset in one location and a private asset in another, but the harvested timber is usually marketed regardless of ownership. The timber, through leasing arrangements, may belong to the private sector even if the forest does not.

While not as obvious as the sale of timber on public lands, the government can also market disposal services—many of which generate significant externalities (such as polluting air or water resources)—by establishing permits and fees. Without fees, the value of waste disposal services provided by a water body or an air shed typically will not be fully reflected in the conventional economic accounts. The extent to which it is reflected depends on who bears the costs associated with the disposal—that is, on whether the created externalities affect only other market producers (in which case it will be reflected in their input costs or output values) or if it impacts valued nonmarket activities or assets (in which case it will not be reflected). We return to this issue below.

Clear rights of ownership, whether by private or public entities, do not guarantee coverage in conventional economic accounts. Regardless of ownership, a particular natural asset may generate a number of different goods and services, some of which are marketed and some of which are not. In our timber example, one could imagine that, if the nonmarket (non-timber) services—such as its contribution to watershed storage, its support of flora and fauna, or its support of recreation—were measured in monetary terms, the asset and output value of a forest could be many times the value determined by timber sales alone. One of the objectives of nonmarket accounting is to capture these kinds of output values.

Given the often complex ownership arrangements for natural and environmental wealth, and that some of the generated goods and services are already captured in conventional accounts, it is not surprising that various accounting schemes differ in their coverage. Some cover only nonmarket activity that is excluded from the conventional accounts: natural resource depreciation, nonmarketed goods and services, and pollution. Often the intent of these schemes is to augment the conventional accounts for the purpose of generating a superior ("greener") measure of gross domestic product (GDP). Others have chosen a more comprehensive approach, covering not only nonmarket activity, but also market items (such as timber and mineral sales and the costs of pollution reduction) that are already covered in the conventional accounts.

As set out elsewhere in this report, the panel favors the development of satellite accounts that include both the market and the nonmarket components of the activity in question. For the environment, an inclusive approach would more fully highlight the economic importance of the environment and natural resources and serve to disaggregate market activity so as to identify costs and expenditures that are related to the environment and to its protection.[2] To reiterate, however, comprehensive sectoral accounts cannot simply be added up, either with each other or with the conventional income and product accounts—perhaps with the intent of generating a green GDP—since there would be double-counting.

Subsoil Resources

Currently, the NIPAs capture the value of minerals and other subsoil resources that are extracted and sold in the market economy. The NIPAs do not reflect changes in the stock of nonreproducible (or extremely slowly reproducible) assets such as oil or mineral reserves (Nordhaus, 2004, p. 13). Changes in these assets are excluded, not because they have no value, but because they have not yet flowed through markets. This is not unlike the current treatment of changes in the stocks of human or health capital explored in Chapters 4 and 5.

The BEA's expanded IEESA effort produced a "full and well-documented" set of subsoil mineral accounts that did value reserves (National Research Council, 1999, p. 4). The IEESA accounting of net investment in subsoil assets includes a negative entry for depletions or extractions of subsoil assets and a positive entry for additions to reserves. The IEESA methodology is similar to the NIPA treatment of conventional capital formation. We note that this approach is not universally accepted: the SEEA, for example, does not account for new discovery of minerals; only negative changes in the value of subsoil assets are possible in that account.

We support the IEESA approach to accounting for nonreproducible asset

[2]This task can be very difficult since environmental expenditures are often hard to separate from nonenvironmental expenditures. A firm may buy new equipment for reasons of profitability. Coincidentally, this equipment may be "cleaner." In addition, some environmental control costs are not reflected in any increased expenditures. For example, a paper company may choose to reduce pollution simply by eliminating the production of bright paper. This choice has real economic costs for the firm, but it is not reflected in any increases in the costs of pollution-control equipment.

reserves, as it is more analogous to the sensible way in which assets are treated elsewhere in the accounts. Cast in an input-output framework, the inputs reflect the net investment in subsoil resources, primarily spending on resource discovery and extraction. The output side reflects changes in the stock of reserves, valued at current prices, that have not yet been marketed. This calculation is in addition to the value of extractions that are sold in markets already captured as output in the NIPAs.

Although natural resource discovery and depreciation are not reflected in conventional economic accounts and while attempts to do so involve an interesting controversy, accounting for these activities would not represent nearly so quantitatively significant a change as most of the others discussed in this report. This is because, empirically, resource discoveries and resource depletions are largely offsetting. Furthermore, because of their near-market character, including discovery and depreciation in national economic accounts (conventional or satellite) has been considered one of the least difficult nonmarket accounting recommendations to implement. Indeed, methods for estimating depreciation for both mines and forests exist (and are discussed in *Nature's Numbers*) and have been used. This is not to say that asset depreciation is completely straightforward in practice. Even if there is only one product generated by an asset, such as a particular mineral, determining its natural lifetime is tricky. The gross stock of a mine is often uncertain and can even be hard to define. Similarly, in valuing discoveries of additional reserves, minerals that cannot be extracted economically do not have an equivalent value to those that can. That mines have been shut down due to lack of economic viability and then reopened after price increases, is indicative of this phenomenon. Ideally, proved and unproved (and easy and difficult to exploit) reserves should be priced differently (see Nordhaus, 2004, p. 15).

Renewable Environmental Resources

In terms of the magnitude of unmeasured nonmarket activity, renewable environmental resources and pollution are much more significant than subsoil assets. Their quantification and valuation involve more difficult methodology and data issues as well. In part because of this difficulty, BEA's IEESA program had not yet begun to develop this part of the account at the time of its termination.

The most interesting accounting issues relate to air and water pollution, for which externalities carry potentially very high values. The value of goods and services that can be produced from environmental resources are clearly linked to changes in the level of pollution. Some of the effects of pollution are captured in the market accounts, but some are not. For example, a reduction in the amount of sulfur dioxide (SO_2) emissions or ground-level ozone may result in reduced worker absence due to illness (a market effect) but also in nonpriced health gains. Likewise, factors contributing to the quality of the environment may or may not be manifest in market expenditures. The cost of catalytic converters is directly reflected in automobile prices, but a firm's choice of production technology may not fully embody the costs of pollution associated with different options.

The extent to which the effects of pollution are captured in the NIPAs depends on who bears the costs. Nordhaus (2004) points out that there are two relevant cases involving externalities. In the first case, the entire impact of an externality flow is reflected in the market accounts, even though no market transaction occurs. If a chemical firm pollutes a nearby water source, and the sole harm that arises from that action is that a farmer's crop yield (sold at market) is reduced, the flow takes place within the market. For accounting purposes, this case is a concern only if we want to disaggregate production accurately by sector—here, chemicals and agriculture. The second case, in which externalities flow across the market boundary, is important for aggregate output measurement. If pollution from the chemical plant impinges on nonmarket recreational opportunities or the population's health status, then failure to account for these effects will distort total output and welfare measures (Nordhaus, 2004, p. 8). Standard accounting methods in the NIPAs would need revision to account properly for the second case since externality disaggregation changes the value added in both the nonmarket and market sectors.

The relevance of this kind of information to policy is fairly obvious. Accounting data on externalities would assist policy makers charged with setting taxes or permit fees for emissions of pollutants or disposal of industrial waste. While market incentives are not yet widely used in environmental regulation in the United States, there is precedent; one important example is the marketable permits used in the acid rain trading program. If properly set, taxes and fees closely approximate the costs of damage associated with a harmful activity, thereby encouraging socially optimal decisions about production processes. But, whether or not fees and charges reflect the true value[3] of the air and water services provided (such as the positive value of waste disposal services or the negative value of the pollution associated with waste disposal), a society that charges firms for the right to pollute will, by conventional market measures, look different from an otherwise similar society that is laissez faire regarding externalities.

---

[3]The term "value" in this chapter refers to the amount individuals in society would be willing to pay for the services of the environment and to avoid any associated detrimental effects such as pollution. This valuation concept takes as given that individuals are the proper arbiters of value and that the environmental services (or damages) can, in principle, be substituted for equivalent amounts of other goods and services (including money) such that the level of value prior to the substitution can remain unchanged. These two assumptions are at the core of the concept of economic value as developed in the modern theory of neoclassical welfare economics; however, "economic value" may not be equivalent to other philosophical concepts of value such as intrinsic value or religious value. For an excellent discussion of economic value and how the concept relates to the techniques of benefit and welfare measurement, see Freeman (1993).

In an accounting framework, there are two ways to handle environmental improvement or degradation that is tied to market production. One can think of pollution created by a firm in the course of its production of goods as a negatively valued output—the firm is producing goods, but it is also producing harmful emissions. Also, one can think of pollution-related environmental damage as a cost of production—to produce, the firm needs workers, equipment, and the environment for waste disposal. In either case, pollution can in many cases be quantified in terms of particulate levels or other physical units. It should be noted that pollution damage and the input of waste disposal services are not alternative measures of exactly the same thing: pollution associated with production. In fact, they are usually unequal in dollar terms and, indeed, waste disposal values can be quite high even when pollution damage is near zero (the reverse is also possible). For this reason, environmental accounting systems should keep these concepts distinct. Valuing degradation as it affects nonmarket outputs (e.g., health and recreation) is difficult because the link between pollution and health is not well understood and because valuing health increments is controversial, though the development of such valuations is a clear goal. Ideally, methods for valuing changes in output associated with changes in the quality of the environment would parallel those developed in the other nonmarket accounts.

CURRENT ACCOUNTING APPROACHES

Formal accounting systems, such as the System of National Accounts (SNA) or the NIPAs, provide a useful way of viewing the interactions between the inputs and outputs that characterize the modern economy. Indeed, one definition of a formal accounting system is that it is a summary of a *process*—a relationship that transforms inputs into outputs. Ideally, the accounting structures proposed and described in this report would contain information that would help researchers better understand underlying production functions for areas of nonmarket activity.

A principal weakness of conventional accounting systems is that they ignore those inputs and outputs that do not trade in ordinary markets. In other words, they misspecify the economy's "true" production function. Nonmarket accounting can be viewed as an effort to correct this misspecification. Concerns with the particular failure to measure nonmarketed environmental inputs and outputs has a history that parallels the concern with the failure of conventional economic output measures to reflect deteriorating environmental quality.[4]

There is widespread agreement that the conventional accounting systems, in spite of their neglect of environmental factors, have served their users well. Thus,

[4]Leontief (1970) is an important early article on supplemental input-output accounting for the environment.

most environmental accounting systems are designed as satellite accounts, as recommended throughout this report for other nonmarket areas. Still, environmental accounting efforts can be developed in close proximity with the conventional economic accounts. For example, the ENRAP system developed for the Philippines augments the conventional production inputs by including environmental waste-disposal services and the conventional production outputs by including recreation, esthetic, and biological support services, as well as the negative outputs associated with pollution (Peskin and delos Angeles, 2001). Thus, ENRAP has the appearance of an SNA-type input-output accounting structure with more rows and columns.

While certain systems attempt to maintain close consistency with the SNA (with respect, for example, to sector definitions or valuation measures), the reliance on satellite accounts lessens the need for such consistency. Thus, for example, the Dutch NAMEA system (as well as the early Leontief system) relies on physical measures rather than the monetary valuations of the conventional accounts (Keuning, 1995). The Philippine ENRAP system relies on monetary measures, but its valuations often include consumer and producer surplus and are thus inconsistent with the market-type valuations of the SNA. The proponents of ENRAP do not consider this lack of consistency with the conventional accounts a major defect since the deficiency is driven by the type of additional information needed by users of the system.

The proponents of other systems, most notably the U.N.'s SEEA, appear more concerned with consistency with the conventional SNA, but the accounting consistency that has been attained with SEEA is not costless. The SEEA 2003 comprises four major components: flow accounts for pollution, energy, and materials; environmental protection and resource management expenditure accounts; natural resource asset accounts; and valuation of nonmarket flow and environmentally adjusted aggregates.[5] In an effort to maintain exact SNA definitions of productive sectors, SEEA cannot cover nonmarketed productive services of the natural environment, such as those associated with recreation, flora and fauna support, or waste disposal. In addition, current implementations of SEEA measure environmental damage and resource depletion by the cost of restoration. While the authors of SEEA recognize that such valuations are not justified in theory, the use of cost valuations is consistent with most SNA practice; for example, the SNA measures the value of governmental services at their cost. As a result, however, SEEA data cannot be used to assess the efficiency of prospective policies to protect the environment. The benefits of such policies are, by definition, equal to their cost.

In spite of its weaknesses, SEEA has provided the framework for many

---

[5]Full details of these accounts can be found at: http://unstats.un.org/unsd/environment/seea2003.pdf [accessed October 14, 2004].

environmental accounting efforts worldwide. The suspended U.S. effort within the BEA, the IEESA system, was also an extension of the SNA (Bureau of Economic Analysis, 1994a). The overall structures and scope of SEEA and IEESA are similar. Both systems neglect nonmarket services of the natural environment. Perhaps the most important difference between them is that SEEA does not consider mineral discoveries as additions to capital stock (based on the argument that the minerals were already there in nature—they were simply unknown to man), while IEESA does count them.

FUTURE DIRECTIONS

For environmental accounting, broad agreement on approach or on implementation techniques has not yet emerged—again, in part because different users require different kinds of information. Experience has shown that it is technically easier to develop accounts for specific natural-resource sectors, such as mining and forestry, than for more general areas of environmental concern, such as air and water quality. As a result, many environmental accounting efforts have taken a staged approach, concentrating first on sector-by-sector natural-resource accounting and then, if financial resources permit, shifting focus to more general areas of environmental quality.

While *Nature's Numbers* (National Research Council, 1999) was sympathetic to this staged approach, it nevertheless recommended a more comprehensive approach: directing the effort to all areas that may be of environmental significance. The problem with the staged approach is that efforts tend to be directed towards problems that are easier and not necessarily those that are of more policy significance.

The comprehensive approach is, of course, not without problems as well. First, it requires coordination. Ideally, there should be a commitment to a common framework but, as noted, there is no general agreement as to which of the several alternative environmental accounting approaches used around the world is the best. While *Nature's Numbers* was critical of some features of certain accounting approaches, it made no recommendations regarding the best choice of accounting system. *Nature's Numbers* did strongly back elements of the IEESA—for example, supporting the use of monetary (or valuation) systems, as opposed to the physical accounting approaches. Second, a comprehensive approach can make great demands on available expertise and accounting resources. Choice of approach may also involve tradeoffs among desirable attributes, such as accounting accuracy, policy relevance, and consistency with market accounts.

Estimation Approaches and Practice

While *Nature's Numbers* was not intended to be a handbook on environmental accounting techniques, it did provide a general discussion of estimation methods, with more detail regarding estimates in the minerals and forestry sectors. This panel very much agrees with that book's conclusion that, to the full extent possible, monetary estimates should rely on observed market prices and actual market behavior, rather than surveys projecting human behavior (e.g., contingent valuation studies). In addition, *Nature's Numbers* expressed a preference for marginal valuations as opposed to total valuations that include consumers' surplus. A major reason for this preference is to ensure that natural and nonmarket capital is treated in a manner consistent with the market-oriented NIPAs, in part so that valuations obtained for the former do not dwarf the value of capital for market sectors (Nordhaus, 2004, p. 13).

The *Nature's Numbers* panel recognized, however, that many existing valuation techniques, such as the travel-cost method to value nonmarket recreation services, provide estimates that contain consumer surplus. In actual practice, budget, staffing, and the lack of available data limit the ability to undertake de novo data development. Experience has demonstrated that making just one nonmarket estimate—such as the monetary value of a particular recreation area or the impact in monetary terms of water pollution in the Chesapeake Bay—can cost many thousands of dollars (Grambsch et al., 1993). Imputing the many similar estimates that would be required for a full set of environmental accounts would quickly exhaust many existing research budgets. Thus, the usual practice has been to "borrow" estimates made elsewhere or for other purposes. For example, an important source has been analyses undertaken by the U.S. Environmental Protection Agency (EPA) in connection with the development of its air and water regulations. The procedure for using these estimates requires an appeal to assumptions and a reliance on economic theory. The process can be illustrated with the ENRAP environmental accounts.

An important entry on the "input" side of the ENRAP accounts is an estimate of the value of waste-disposal services provided by the environment to various industrial sectors (as well as agriculture, households, and governments). In the ENRAP system, this value is intended to equal what these sectors would be willing to pay for these services. In the absence of direct observations of willingness to pay, inputs were measured by estimating what the cost to a particular sector would be were it denied access to the waste-disposal service. These costs, in turn, were approximated by EPA data on the costs of meeting pollution abatement regulations.

A number of assumptions are involved when accepting that the EPA estimates approximate the intended willingness-to-pay measure. For example, one must assume that the EPA cost estimates are efficient—that they represent the least-cost way to reduce pollution. In addition, one must also assume that the regulations (and the accompanying cost estimates) actually would eliminate all uses of the environment for waste-disposal purposes. In fact, many of the regulations are more practical; often the target is about 90 percent reduction from current levels—not the 100 percent that would be consistent with the intended willingness-to-pay measure. As discussed in Breyer (1993), the marginal cost of cleaning up "the last 10 percent" (e.g., from hazardous waste sites) can be very large.

With this reliance on EPA data, ENRAP developers were able to confine new estimation to a minority of sectors for which it was believed that the EPA estimates were either totally inappropriate or simply not available. This practice greatly reduced the overall developmental costs. Nevertheless, the cost of the entire accounting process was not trivial. The Philippines ENRAP accounts, for example, cost about $4 million. This figure did cover several illustrative policy analyses considered important to the account development effort.

When practical concerns dictate some reliance on borrowed data, it becomes unlikely that a final set of environmental accounts will meet ideal objectives. Data gaps and inconsistencies (especially with the conventional income and product accounts) should be expected. An important question is whether, in the presence of these deficiencies, the nonmarket accounting effort is still worthwhile. Past nonmarket environmental accounting efforts have addressed this question by considering whether the additional information—albeit imperfect—make the benefits of nonmarket accounting worth its costs. The answer depends largely on the purpose motivating the effort. At least with respect to environmental accounting, if the purpose is merely to obtain a measure of an environmentally adjusted GDP, there is, indeed, some question of whether the effort is worthwhile. Not only is it very difficult to obtain ideal measures, the less-than-perfect adjustments to GDP seem relatively small (though this may not be the case with other nonmarket activities covered in this report). Nevertheless, the information content of the satellite environmental accounts would be valuable in managing the nation's assets and for improving regulatory decisions. As concluded in *Nature's Numbers*: "improved natural-resource and environmental accounts can provide useful information on natural assets under federal management. . . . In the case of environmental resources such as air and water quality, a comprehensive set of environmental accounts would provide useful information on the economic returns the nation is reaping from its environmental investments" (National Research Council, pp. 31-32). One major finding in the environmental area concerns the relative importance of industrial versus household and agriculture pollution. Although policy has focused on the former, the environmental accounts have suggested that, for many pollutants, the major problem lies with the latter two sectors.

Linkage with Other Nonmarket Accounting Efforts

Efforts to build satellite environmental and natural-resource accounts have not typically been carried out alongside similar efforts in other areas of nonmarket accounting. The ENRAP effort was an exception since it originally was part of the Measurement of Economic and Social Performance project (MESP) funded by the National Science Foundation (National Bureau of Economic Research,

1978). Since the vast majority of environmental accounting efforts have been independent projects, there is little experience regarding how these efforts would have been affected were they part of a larger nonmarket accounting program.

One can speculate, however, that, if work were progressing in other areas of nonmarket accounting, useful exchanges of information and methodology might occur. For example, estimates of the value of changes in morbidity and mortality developed in connection with health accounts could provide useful input into estimates of the value of those changes that are due specifically to changes in pollution levels.[6] Such estimates have been used to measure the willingness to pay to reduce pollution—an important entry in a number of environmental accounting systems. Of course, information can flow in both directions. Those developing health accounts would probably find it useful to inspect the many estimates of mortality and morbidity that can be found in the environmental quality literature.

Besides exchanges of data and information, one could promote coordinated research on valuation methods. The various techniques for valuing nonmarket services of the environment and associated damages that result from use of these services—hedonic estimation, travel cost methods, and the direct willingness-to-pay estimates typically used to value mortality and morbidity—could play a role in nonmarket accounting generally. In fact, one could imagine joint estimation efforts. For example, a well-designed property value study could support the development of those partial price elasticities needed to value the benefits of improved schools, better policing, improved infrastructure, and cleaner air and water—factors that not only help to determine the value of property, but also are nonmarket outputs of the government and environmental sectors. Of course, whether such joint estimation is efficient largely will depend on the available data. Estimating the various parameters associated with the above outputs assumes a rich data set free of strong collinearities. Lacking the variation in the data required to identify the parameters of interest, it might be more efficient to develop data specific to the parameter of interest. Thus, the effect of air pollution might be ascertained by looking at property values in regions that demonstrate sharp differences in air quality but not necessarily in school quality, while the effect of schools might be studied with data from areas that had sharp differences in school quality but not necessarily in air quality.

Data Needs

While the accounting objective requires monetary valuations, *Nature's Numbers* did discuss the need for more complete physical information, especially data

[6]See National Research Council (2002b), which summarizes research that estimates the public health benefits of proposed air pollution regulations.

relating to environmental quality and its possible effects on human health and other activity. The emphasis on the need for better physical information reflects a reality familiar to those undertaking environmental benefit-cost analysis: knowledge about the methods of monetary valuation of environmental changes often exceeds knowledge of the physical changes themselves. For example, there are techniques that allow valuation of adverse health effects due to air and water pollution. Far less is known about the connection between any particular measured ambient air and water quality parameter and how this level of quality may affect human populations. Not only is there uncertainty as to effects on particular individuals, it is not clear which individuals are actually exposed.

Indeed, data on actual ambient conditions can be very spotty. Water quality data, for example, are collected in relatively few locations in a timely and consistent manner, even though conditions can vary tremendously over time due to changes in water flows, temperature, and pollution discharges. Air pollution measurement also can be very spotty, especially in non-urban areas. Information on the stock, growth, and depletion of natural resources may be relatively more complete, but *Nature's Numbers* cited needs for improvement even in this area, especially for resources under federal domain and for fisheries.

Since much of the physical information needed to develop environmental and natural resource accounts is collected by various federal agencies as part of their administrative functions, cooperation with these agencies is essential. *Nature's Numbers* (National Research Council, 1999, pp. 196-201) contains a useful appendix describing sources of physical and valuation data on natural resources and the environment. It includes details on the scope of activities and resources valued, and on valuation methods used; it also provides a good starting place for establishing where new data are needed.

A NOTE ON THE SOCIAL ENVIRONMENT

In this chapter we have briefly described how environmental resources factor into productive activities, whether market oriented or not. The environment, however, consists of not just physical factors, but social ones as well, and the social environment can affect the population's productivity and well-being in ways that are at times analogous to influences of the physical environment. What is the social environment? People's social networks consist of family, friends, and community. As discussed in Chapter 4, social factors influence the effectiveness with which families can develop human capital in children. This is very obvious, for example, in populations where household violence is prevalent. We also observe the importance of safety and level of support in school environments in the production of educated individuals, the subject of Chapter 5.

While significant progress has been made in the area of physical environmental accounting, comparatively little has been done on the social side. One possible reason that environmental accounting research has generally neglected the social environment is that some aspects of it, while not lacking value per se, lack economic value in the sense that they are scarce. Justification for omitting social factors from environmental accounting might follow the reasoning for ignoring, say, energy production from the sun, which is valuable but not economically scarce. In this report, we do not address the social environment in detail, but we acknowledge that a comprehensive set of satellite nonmarket accounts would include information on the social environment, covering such factors as the degree of social cohesion, the stability and effectiveness of political systems, and levels of safety and security. An account ideally would link these largely nonmarket inputs to various market and nonmarket output aggregates.

The social environment produces an output commonly referred to as social capital (the term is defined and briefly discussed in Chapter 1). Often, social capital is evaluated at specific geographic levels of detail, such as the community. One can observe many community-level variables that might relate to measures of social capital—such as membership in local organizations, volunteering, crime, and such public goods as education and local environmental amenities. But the quantity of social capital, as embedded in these indicators, is hard to quantify, let alone to price.

One dimension of the social environment for which data and concepts are relatively developed is in the area of crime. For crime, one can easily observe quantities, though not prices (a feature common to many local public goods). There are four basic approaches that could be used to attempt to account for crime. One would be to take all of the income and product in the conventional NIPAs that relate to security systems, alarms, locks, etc., and subtract that out from standard market accounting data. Unfortunately, this tells us little if anything about how societal well-being changes as crime rises or falls. Another option would be to look at changes in murder rates and use value-of-life estimates as the price in a way that parallels valuation in health. The problem with this approach is that it helps us with only one, albeit an important, class of crime.

An encouraging fact about local public goods is that they should be reflected in housing prices. We have discussed how housing prices might be used to help value the output of schools and even aspects of the local physical environment. This observation suggests two further approaches for accounting for crime: use of hedonic methods to value willingness to pay for reductions in crime; or of use discrete choice econometric models—in which people are presumed to choose from a discrete and finite set of options—to estimate that willingness to pay.

Hedonic techniques have been applied by Smith and Huang (1995), Chay and Greenstone (2004), and Banzhaf (2002) to examine the value of air quality. Papers that have used housing hedonics to estimate quality-of-life indexes include Blomquist et al. (1988) and Kahn (1995). Banzhaf (2002) incorporated air quality, crime, and education in his index for Los Angeles for the period 1989-1994. Chattopadhyay (1999) and Palmquist and Israngkura (1999) estimated the value of air quality using discrete choice models. Banzhaf (2002) used discrete choice models to estimate the value of air quality, crime, and education in Los Angeles.

There are still many problems with these methods, such as how to aggregate across households (see Pollak, 1989, on social Laspeyres indexes). Relative to hedonics, the discrete choice estimation approach has the advantage of allowing for heterogeneity in tastes and for allowing the "filling in" of the product space with new alternatives. But hedonic methods are computationally more tractable, and statistical agencies have had experience applying them. While we do not yet know whether in practice the two approaches would consistently yield similar results, Banzhaf (2002) found that the percent change in the cost of living implied by the results from his hedonic regressions were similar to those from his discrete choice models.

The hedonic and discrete choice models raise some of the same issues. Both require that all public goods be included. Otherwise, an estimate of the cost of living is still biased. Both models also would require frequent estimation. Not only are quantities changing over time, so too are relative prices. For example, Cragg and Kahn (1997) find that people's willingness to pay for a temperate climate has increased over time. Costa and Kahn (2003) find that value of life has increased over time. Most valuation work does not take price changes into account (e.g., Chay and Greenstone, 2004), but these can be very important and at least in the methodologies outlined above there is nothing that prevents doing so. Both hedonic and discrete choice models also raise the issue of what is a neighborhood. What level of detail in terms of geography and household characteristics do we need to work with to obtain good results? These are all areas for further research.

References

Abowd, J.M., J. Haltiwanger, and J. Lane
2004 Integrated longitudinal employee-employer data for the United States. *American Economic Review* 92(2):224-229.

Acemoglu, D., and J.D. Angrist
2001 How large are human capital externalities? Evidence from compulsory schooling laws. In *NBER Macroeconomic Annual 2000,* B.S. Bernanke and K. Rogoff, eds. Cambridge, MA: MIT Press.

Aguiar, M., and E. Hurst
2004 *Consumption vs. Expenditure*. NBER Working Paper No. 10307. Cambridge, MA: National Bureau of Economic Research.

Ashenfelter, O., C. Harmon, and H. Osterbeek
1999 A review of estimates of the schooling/earnings relationship, with tests for publication bias. *Labour Economics* 6(4):453-470.

Atkinson, T.
2004 *Atkinson Review: Interim Report. Measurement of Government Output and Productivity for the National Accounts.* Available: http://www.statistics.gov.uk/about/methodology_by_theme/atkinson/downloads/atkinson.pdf [accessed October 12, 2004].

Aulin-Ahmavaara, P.
1991 Production prices of human capital and human time. *Economic Systems Research* 3(4):345-365.
2003 The SNA93 values as a consistent framework for productivity measurement: Unsolved issues. *Review of Income and Wealth* 49:117-133.

Australian Bureau of Statistics
2000 *Unpaid Work and the Australian Economy, 1997*. Canberra: Australian Bureau of Statistics.

Banzhaf, H.S
2002 *Quality Adjustment for Spatially-Delineated Public Goods: Theory and Application to Cost-of-Living Indices in Los Angeles*. Discussion Paper 02-10. Washington, DC: Resources for the Future.

Becker, G.
1965 A theory of the allocation of time. *Economic Journal* 49(299):493-508.
1973 A theory of marriage: Part I. *Journal of Political Economy* 81(4):813-846.
1974 A theory of marriage: Part II. *Journal of Political Economy* 82(2):S11-S26.

Ben-Porath, Y.
1980 The F-connection: Families, friends, and firms and the organization of exchange. *Population and Development Review* 6(March):1-30.

Berndt, E.R., D.M. Cutler, R.G. Frank, Z. Griliches, and J.E. Triplett
2000 Medical care prices and output. In *Handbook of Health Economics*, A.J. Culyer and J.P. Newhouse, eds. Amsterdam, The Netherlands: Elsevier.

Bianchi, S.M.
2000 Maternal employment and time with children: Dramatic change or surprising continuity? *Demography* 37(November):139-154.

Bianchi, S., P.N. Cohen, S. Raley, and K. Nomaguchi
1985 Inequality in parental investment in child-rearing: Expenditures, time and health. In *Social Inequality*, K. Neckerman, ed. New York: Russell Sage.

Bianchi, S.M., J.P. Robinson, and L.C. Sayer
2001 *Family Interaction, Social Capital, and Trends in Time Use Study (FISCT).* Ann Arbor, MI: Inter-University Consortium for Political and Social Research (ICPSR).

Black, S.
1999 Do better schools matter? Parental valuation of elementary education. *Quarterly Journal of Economics* 114(2):577-599.

Blau, F., and L. Kahn
2001 *Do Cognitive Test Scores Explain Higher US Wage Inequality.* NBER Working Paper No. 8210. Cambridge, MA: National Bureau of Economic Research.

Blomquist, G.C., M.C. Berge, and J.P. Hoehn
1988 New estimates of quality of life in urban areas. *American Economic Review* 78:89-107.

Bogart, W., and B. Cromwell
1997 How much is a good school district worth? *National Tax Journal* 50(June):215-232.

Boris, Elizabet T.,
1998 *Myths About the Nonprofit Sector.* Washington, DC: The Urban Institute. Available: http://www.urban.org/url.cfm?ID=307554 [accessed September 17, 2004].

Boskin, M.J., E. Dulberger, R. Gordon, Z. Griliches, and D. Jorgenson
1996 *Toward a More Accurate Measure of the Cost of Living*. Final Report to the U.S. Senate Finance Committee. Washington, DC: Commission to Study the Consumer Price Index.

Bowles, S., and H. Gintis
1977 *Schooling in Capitalist America*. New York: Basic Books.
2000 Does schooling raise earnings by making people smarter? In *Meritocracy and Economic Inequality*, K. Arrow, S. Bowles, and S. Durlauf, eds. Princeton, NJ: Princeton University Press.

Bowles, S., H. Gintis, and M. Osborne
2001 The determinants of earnings: A behavioral approach. *Journal of Economic Literature* 39(4):1137-1176.

Breyer, S.
1993 *Breaking the Vicious Circle: Toward Effective Risk Regulation.* Cambridge, MA: Harvard University Press.

Brooks-Gunn, J. G.J. Duncan, P. Kato Klebanov, and N. Sealand
1993 Do neighborhoods influence child and adolescent development? *American Journal of Sociology* 99(2):353-395.

Budig, M.J., and N. Folbre
2004 Activity, proximity, or responsibility? Measuring parental childcare time. In *Family Time: The Social Organization of Care*, N. Folbre and M. Bittman, eds. New York: Routledge.

Bureau of Economic Analysis

1985 *An Introduction to National Economic Accounting.* U.S. National Income and Product Accounts, MP-1. Washington, DC: U.S. Department of Commerce.

1994a Integrated economic and environmental satellite accounts. *Survey of Current Business* (April):33-49.

1994b A satellite account for research and development. *Survey of Current Business* (November):1-26.

1995 Mid-decade strategic review of BEA's economic accounts: An update. *Survey of Current Business* 75(April):48-56.

2004 National income and product account tables. *Survey of Current Business* (August):30-166.

Cabrera, N.J., C. Tamis-LeMonda, R. Bradley, S. Hofferth, and M. Lamb

2000 Fatherhood in the 21st century. *Child Development* 71(1):127-136.

Card, D.

1999 The causal effect of education on earnings. In *Handbook of Labor Economics*, Volume III, O. Ashenfelter and D. Card, eds. Amsterdam, The Nerthlands: Elsevier.

Carneiro, P., and J.J. Heckman

2002 *The Evidence on Credit Constraints in Post-Secondary Schooling.* NBER Working Paper No. 9055. Cambridge, MA: National Bureau of Economic Research.

2003 *Human Capital Policy.* IZA Discussion Paper No. 821. Available: http://ssrn.com/abstract=434544 [accessed November 9, 2004].

Carneiro, P., F. Cunha, and J. Heckman

2003 Interpreting the Evidence of Family Influence on Child Development. Paper presented at The Economics of Early Childhood Development: Lessons for Economic Policy Conference, The Federal Reserve Bank of Minneapolis and The McKnight Foundation in cooperation with the University of Minnesota. October 17, 2003. Department of Economics, University College, London.

Carson, C.

1975 History of the United States National Income and Product Accounts: The development of an analytical tool. *Review of Income and Wealth* 23(June):153-181.

Case, A., D. Lubotsky, and C. Paxson

2002 Economic status and health in childhood: The origins of the gradient. *American Economic Review* 92(5):1308-1334.

Chattopadhyay, S.

1999 Estimating the demand for air quality: New evidence based on the Chicago housing market. *Land Economics* 75(1):22-38.

Chay, K., and M. Greenstone

2004 Does Air Quality Matter? Evidence from the Housing Market. Unpublished paper. Department of Economics, University of Chicago.

Cohen, D., and M. Soto

2002 *Why are Poor Countries Poor? A Message of Hope which Involves the Resolution of a Becker/Lucas Paradox.* CEPR Discussion Paper No. 3528. London, U.K.: Centre for Economic Policy Research.

Coleman, J.S., E.Q. Campbell, C.J. Hobson, J. McPartland, A.M. Mood, F.D Weinfeld, and R.L. York

1966 *Equality of Educational Opportunity.* Washington, DC: U.S. Government Printing Office.

Commission of the European Communities, International Monetary Fund, Organisation for Economic Co-operation and Development, United Nations, and World Bank

1994 *System of National Accounts 1993.* Paris: Organisation for Economic Co-operation and Development.

Conference on Research in Income and Wealth
1958 *A Critique of the United States Income and Product Accounts.* Studies in Income and Wealth, Volume 22. National Bureau of Economic Research. Princeton, NJ: Princeton University Press.

Costa, D.L.
2003 *Health and Labor Force Participation over the Life Cycle: Evidence from the Past.* National Bureau of Economic Research Conference Report. Cambridge, MA: National Bureau of Economic Research.

Costa, D., and M. Kahn
2003 The rising price of nonmarket goods. *American Economic Review* 93(2):227-232.

Council of Economic Advisers
2003 *Economic Indicators, March 2003.* Washington, DC: U.S. Government Printing Office.

Cragg, M, and M. Kahn
1997 New estimates of climate demand: Evidence from migration. *Journal of Urban Economics* 42:261-284.

Currie, J., and D. Thomas
1999 *Early Test Scores, Socioeconomic Status and Future Outcomes.* NBER Working Paper No. 6943. Cambridge, MA: National Bureau of Economic Research.

Cutler, D.
2004 *Your Money or Your Life: Strong Medicine for America's Health Care System.* Cary, NC: Oxford University Press, Inc.

Cutler, D., and E. Richardson
1997 Measuring the health of the United States population. *Brookings Papers on Economic Activity, Microeconomics,* 217-272.

Cutler, D.M., M. McClellan, J.P. Newhouse, and D. Remler
1998 Are medical prices declining? Evidence from heart attack treatments. *Quarterly Journal of Economics* 113(4):991-1024.

Dow, G., and F.T. Juster
1985 Goods, time, and well-being: The joint dependence problem. In *Time, Goods and Well-Being,* F.T. Juster and F. Stafford, eds. Ann Arbor, MI: Institute for Social Research.

Dreze, J., and N. Stern
1987 The theory of cost-benefit analysis. In *Handbook of Public Economics,* Vol. II. A.J. Auerbach and M. Feldstein, eds. Amsterdam, The Netherlands: North-Holland.

Duncan, G.J., and R. Dunifon
1998a 'Soft-skills' and long-run market success. In *Research in Labor Economics, Vol. 17,* S.W. Polachek, ed. Stamford, CT: JAI Press.
1998b Long-run effects of motivation on labor-market success. *Social Psychology Quarterly* 61(1):33-48.

Duncan, G.J, R. Dunifon, and D. Knutson
1996 *Vim Will Win: Long Run Effects of Motivation and Other Non-Cognitive Traits on Success.* IPR Working Paper 96 23. Department of Economics, Northwestern University.

Durlauf, S.
2004 Neighborhood effects. In *Handbook of Urban and Regional Economics,* Vol. 14, Geography and Cities. J.V. Henderson and J-F. Thisse, eds. Amsterdam: North-Holland.

Eeckhoudt, L.R., and J.K. Hammitt
2001 Background risks and the value of a statistical life. *Journal of Risk and Uncertainty* 23(3):261-279.

Egerton, M., K. Fisher, and J. Gershuny
2004 Preliminary Report: Preparing a Harmonised U.S. Heritage Time-Use Datafile. Unpublished report to the Glaser Foundation. University of Essex (February).

Eisner, R.
1989 *The Total Incomes System of Accounts.* Chicago: University of Chicago Press.

El Serafy, S.
1989 The proper calculation of income for depletable natural resources. In *Environmental Accounting for Sustainable Development*, Y.J. Ahmad, S. El Serafy, and E. Lutz, eds. Washington, DC: The World Bank.

Ellwood, D., and T. Kane
2000 Who is getting a college education? Family background and the growing gaps in enrollment. In *Securing the Future: Investing in Children from Birth to College*, S. Danziger and J. Waldfogel, eds. New York: Russell Sage Foundation.

England, P., and N. Folbre
2000 Reconceptualizing human capital. In *The Management of Durable Relations*, W. Raub and J. Weesie, eds. Amsterdam, The Netherlands: Thela Thesis Publishers.

Farrell, P., and V.R. Fuchs
1982 Schooling and health: The cigarette connection. *Journal of Health Economics* 1(3):217-230.

Ferguson, N.
2003 Why America Outpaces Europe (Clue: the God Factor). *New York Times*, June 8, WE3.

Fisk, D., and D. Forte
1997 The federal productivity measurement program: Final results. *Monthly Labor Review* (May):19-28.

Fixler, D.J., M.B. Reinsdorf, and G.M. Smith
2003 Measuring the services of commerical banks in the NIPAs: Changes in concepts and methods. *Survey of Current Business* (September):33-44.

Fogel, R.W.
1986 Nutrition and the decline in mortality since 1700: Some preliminary findings. In *Long-Term Factors in American Economic Growth*, S.L. Engerman and R.E. Gallman, eds. Chicago: University of Chicago Press.
2004 *The Escape from Hunger and Premature Death, 1700-2100: Europe, America, and the Third World,* London: Cambridge University Press.

Folbre, N.
2004 Valuing Parental Time: Total Expenditures on Children in the United States in 2000. Paper presented at the workshop on Supporting Children: English-Speaking Countries in International Context. Princeton University, January 7-9, 2004. Department of Economics, University of Massachusetts, Amherst.

Folbre, N., and B. Wagman
1993 Counting housework: New estimates of real product in the U.S., 1800-1860. *Journal of Economic History* 53(June):275-288.

Folbre, N., J. Yoon, K. Finoff, and A.S. Fuligni
2004 By What Measure? Family Time Devoted to Children in the U.S. Unpublished paper, Department of Economics, University of Massachusetts, Amherst.

Fraumeni, B.M.
2000 The Output of the Education Sector as Determined by Education's Effect on Lifetime Income. Paper presented at the Brookings Institution's Workshop on Measuring the Output of the Education Sector. Washington, DC, April 7, 2000. Bureau of Economic Analysis, U.S. Department of Commerce.

Fraumeni, B.M., and S. Okubo
2001 Alternative Treatments of Consumer Durables in the National Accounts. Paper prepared for the BEA Advisory Committee Meeting, Washington, DC, May 11, 2004. Bureau of Economic Analysis, U.S. Department of Commerce.

Fraumeni, B.M., M.J. Harper, S.G. Powers, and R.E. Yuskavage
2004 An Integrated BEA/BLS Production Account: A First Step and Theoretical Considerations. Paper presented at the CRIW Architecture for the National Accounts Conference, Washington, DC, April 16-17, 2004. Bureau of Economic Analysis, U.S. Department of Commerce.

Fraumeni, B.M., M.B. Reinsdorf, B.B. Robinson, and M. P. Williams

2004 Real Output Measures for the Education Function of Government: A First Look at Primary and Secondary Education. Presentation at the Public Services Performance Workshop, National Institute of Social and Economic Research, London, March 2. Available: http://www.niesr.ac.uk/aim/docs/1 [accessed August 2, 2004].

Freeman, A.M., III.

1993 *The Measurement of Environmental and Resource Values.* Washington, DC: Resources for the Future.

Garrison, J., and A. Krueger

2004 *A Method for Producing Historical Human Capital Accounts.* Princeton, NJ: Princeton University Press.

Gates, J.

1984 Education and training costs: A measurement framework and estimates for 1965-76. In *Measuring Nonmarket Activity*, J. Peskin, ed. Washington, DC: U.S. Government Printing Office. Pp. 107-135.

Gates, J., and M. Murphy

1982 The use of time: A classification scheme and estimates for 1975-76. In *Measuring Nonmarket Economic Activity*, J. Peskin, ed. Washington, DC: U.S. Government Printing Office.

Gauthier, A.H., T. Smeeding, and F.F. Furstenberg, Jr.

2001 Do We Invest Less Time in Children? Trends in Parental Time in Canada Since the 1970s. Unpublished manuscript. University of Calgary, Calgary, Alberta.

Ginther, D.K., and R.A. Pollak

2004 Family Structure and Children's Educational Outcomes: Blended Families, Stylized Facts, and Descriptive Regressions. *Demography* 41(November):671-696.

Goldin, C.

1986 The female labor force and American economic growth, 1890-1980. In *Long Term Factors in American Economic Growth*, S. Engerman and R. Gallman, eds. Chicago: University of Chicago Press.

Goleman, D.

1995 *Emotional Intelligence*. New York: Bantam Press.

Graham, J.W., and R.H. Webb

1979 Stocks and depreciation of human capital: New evidence from a present-value perspective. *Review of Income and Wealth* 25(2):209-224.

Grambsch, A.E., R.G. Michales, and H.M. Peskin

1993 Taking stock of nature: Environmental accounting for the Chesapeake Bay. In *Toward Improved Accounting for the Environment*, An UNSTAT-World Bank Symposium, E. Lutz, ed. Washington, DC: The World Bank.

Greenwood, J., R. Rogerson, and R. Wright

1995 Household production in real business cycle theory. In *Frontiers of Business Cycle Research*, Thomas Cooley, ed. Princeton, NJ: Princeton University Press.

Griliches, Z.

1992 *Output measurement in the service sectors*. Studies in Income and Wealth, Vol. 56. Z. Griliches, ed. Chicago: University of Chicago Press.

Gronau, R., and D. Hamermesh

2003 *Time vs. Goods: The Value of Measuring Household Production Technologies*. NBER Working Paper No. 9650. Cambridge, MA: National Bureau of Economic Research.

Grossman, M.

1972 On the concept of health capital and the demand for health. *Journal of Political Economy* 80(2):223-255.

1999 *The Human Capital Model of the Demand for Health.* NBER Working Paper No. 7078. Cambridge, MA: National Bureau of Economic Research.

Hamilton, K.
1996 Pollution and pollution abatement in the National Accounts. *Review of Income and Wealth* 42(1):13-34.

Hammitt, J.K.
2003 *A Common Health-Valuation Framework.* Presentation at the Conference Valuing Health Outcomes: An Assessment of Approaches, February 2003, Resources for the Future, Washington, DC.
Available: http://www.rff.org/rff/Events/Valuing-Health/loader.cfm?url=/commonspot/security/getfile.cfm&PageID=5352 [accessed August 2, 2004].

Hansen, K.T., J.J. Heckman, and K.J. Mullen
2004 The effect of schooling and ability on achievement test scores. *Journal of Econometrics* 121(1):30-98.

Hausman, J.
1996 Valuation of New Goods Under Perfect and Imperfect Competition. In *The Economics of New Goods,* Studies in Income and Wealth, Vol. 58. T. Bresnahan and R. Gordon, eds. Chicago: National Bureau of Research.

Haveman, Robert H., and Barbara L. Wolfe
1984 Schooling and economic well-being: The role of nonmarket effects. *Journal of Human Resources* 19(3):377-407.
1995 The determinants of children's attainments: A review of methods and findings. *Journal of Economic Literature* 33(4):1829-1878.

Health Canada
2003 *The Voluntary Health Sector: Looking to the Future of Canadian Health Policy and Research.* Available: http://www.hc-sc.gc.ca/iacb-dgiac/arad-draa/english/rmdd/wpapers/voluntary.html#value [accessed June 17, 2004].

Hecht, J.
2000 *Lessons Learned from Environmental Accounting: Findings from Nine Case Studies.* Washington, DC: IUCN-The World Conservation Union.

Heckman, J.J.
1976 The common structure of statistical models of truncation, sample selection, and limited dependent variables and a simple estimator for such models. *Annals of Economic and Social Measurement* 5:475-492.

Heckman J.J., and P. Klenow
1997 *Human Capital Policy.* Available: http://www.klenow.com/HumanCapital.pdf [accessed August 2, 2004].

Heckman, J.J., and Y. Rubinstein
2001 The importance of noncognitive skills: Lessons from the GED testing program. *American Economic Review* 91(2):145-149.

Heckman, J., and E. Vytlacil
1999 Local instrumental variables and latent variable models for identifying and bounding treatment effects. *Proceedings of the National Academy of Sciences* 96:4730-4734.

Heckman, J., R. Matzkin, and L. Nesheim
2003 *Simulation and Estimation of Nonadditive Hedonic Models.* NBER Working Paper 9895. Cambridge, MA: National Bureau for Economic Research.

Hofferth, S., and J. Sandberg
2001 How American children spend their time. *Journal of Marriage and the Family* 63:295-308.

Holloway, S., S. Short, and S. Templin
2002 *Household Satellite Account (Experimental) Methodology. United Kingdom Office for National Statistics.* Available: http://www.statistics.gov.uk/hhsa/hhsa/resources/file attachments/hhsa.pdf [accessed November 9, 2004].

Horrigan, M., and D. Herz

2005 A study in the process of planning, designing and executing a survey program: The BLS American Time-Use Survey. In *The Economics of Time Use Data*, D. Hamermesh and G. Pfann, eds. Amsterdam: North-Holland.

Hoxby, C.

2003 School choice and school productivity (Or could school choice be a tide that lifts all boats?). In *The Economics of School Choice*, C. Hoxby, ed. Chicago: University of Chicago Press.

Independent Sector and the Urban Institute

2001 *New Nonprofit Almanac and Desk Reference*. Washington, DC: Independent Sector.

2004 *Giving and Volunteering Survey*. Available: http://www.independentsector.org/programs/research/volunteer_time.html#volunteers [accessed September 17, 2004].

Ironmonger, D.S.

1996 Counting outputs, capital inputs and caring labor: Estimating gross household product. *Feminist Economics* 2(Fall):37-64.

2003 There are Only 24 Hours in a Day! Solving the Problematic of Simultaneous Time. Paper presented at the 25th IATUR Conference on Time Use Research, 17-19 September, Brussels. Available: http://www.vub.ac.be/TOR/iatur/abstracts/view-paper.php?id=91 [accessed November 9, 2004].

Jencks, C.

1979 *Who Gets Ahead? The Determinants of Economic Success in America*. New York: Basic Books.

Jenkinson, G.

2003 Measuring the volume of government outputs. Presentation at the OECD National Accounts Experts Meeting, Paris, October 7-10. Office for National Statistics.

Jorgenson D.W., and B.M. Fraumeni

1989 The accumulation of human and nonhuman capital 1948-1984. In *Measurement of Savings, Investment, and Wealth*, Studies in Income and Wealth Volume 52. R.E. Lipsey and H.S. Tice, eds. Chicago: University of Chicago Press.

1992 The output of the education sector. In *Output Measurement in the Service Sectors*, Studies in Income and Wealth, Vol. 56, Z. Griliches, ed. Chicago: University of Chicago Press.

Jorgenson, D.W., and A. Pachon

1983a The accumulation of human and nonhuman wealth. In *The Determinants of National Saving and Wealth*, R. Hemming and R. Modigliani, eds. London: Macmillan Press.

1983b Lifetime income and human capital. In *Human Resources, Employment, and Development*, P. Streeten and H. Maier, eds. London: Macmillan Press.

Juster, F.T.

1978 *The Distribution of Economic Well-Being*. Studies in Income and Wealth No. 41. New York: Columbia University Press.

Juster, F.T., and F. Stafford

1985 *Time, Goods and Well-Being*. Ann Arbor, MI: Institute for Social Research.

Kahn, M.

1995 A revealed preference approach to ranking city quality of life. *Journal of Urban Economics* (38):221-235.

Kahneman, D., E. Diener, and N. Schwarz, eds.

1999 *Well-Being: The Foundations of Hedonic Psychology*. New York: Russell Sage Foundation.

Kane, T., D. Staiger, and G. Samms

2003 School accountability ratings and housing values. *Brookings-Wharton Papers on Urban Affairs* 2003:83-137.

Katz, A.J.

1983 Valuing the Services of Consumer Durables. *Review of Income and Wealth* 29:405-427.

Kendrick, J.W.

1967 Studies in the National Income Accounts. In *Forty-Seventh Annual Report of the National Bureau of Economic Research*. Cambridge, MA: National Bureau of Economic Research.

1974 *The Accounting Treatment of Human Investment Capital*. New York: Columbia University Press.

Kenkel, D.

1990 Consumer health information and the demand for medical care. *Review of Economics and Statistics* 72:587-595.

Keuning, S.J.

1995 An information system for environmental indicators in relation to the national accounts. In *The Value Added of National Accounting*, W. de Vries, G. den Bakker, M. Gircour, S.J. Keuning, and A. Lenson, eds. Amsterdam, The Netherlands: Central Bureau of Statistics.

King, W.I.

1923 *Employment Hours and Earnings in Prosperity and Depression, United States, 1920-1922*. New York: National Bureau of Economic Research.

King, G., J. Christopher, L. Murray, J.A. Salomon, and A. Tandon

1998 Enhancing the validity and cross-cultural comparability of survey research. *American Political Science Review* 98(1):191-207.

Kodrzycki, Y.K.

2002 Education in the 21st century: Meeting the challenges of a changing world. In *Proceedings from the Federal Reserve Bank of Boston's 47th Economic Conference*. Boston: Federal Reserve Bank of Boston.

Krueger, A.

1999 Measuring labor's share. *American Economic Review* 89(2):45-51.

Krueger, A.B., and M. Lindahl

2001 Education for growth: Why and for whom? *Journal of Economic Perspectives* 39(4):1101-1136.

Kuznets, S.

1934 National Income 1929-1932. Senate Document No. 124, 73rd Congress, 2nd Session. Washington, DC: U.S. Government Printing Office.

Landefeld, J.S., and S.L. Howell

1997 Accounting for nonmarketed household production within a national accounts framework. Presentation at the Conference on Time-Use, Nonmarket Work, and Family Well-being. Bureau of Labor Statistics and MacArthur Network on the Family and the Economy Conference, Washington, DC, November 20-21, Bureau of Labor Statistics.

Landefeld, S.J., and S. McCulla

2000 Accounting for nonmarket household production within a national accounts framework. *Review of Income and Wealth* 46(3):289-307.

LaPlante, M.P., C. Harrington, and T. Kang

2002 Estimating paid and unpaid hours of personal assistance services in activities of daily living provided to adults living at home. *Health Services Research* 37(2):397-415.

Lareau, A.

2003 *Unequal Childhood: Class, Race, and Family Life*. Berkeley, CA: University of California Press.

Leibowitz, A.

1974 Home investments in children. *Journal of Political Economy* 82(2):S111-131.

Leontief, W.

1970 Environmental repercussions and the economic structure: An input-output approach. *Review of Economics and Statistics* 52(3):262-271.

Leroux, J., J. Rizzo, and R. Sickles
2003 Health, Mental Health and Labor Productivity. The Role of Self Reporting Bias. Mimeo, Rice University. Available: http://www.ruf.rice.edu/~rsickles/paper/productivity.pdf [accessed September 17, 2004].

Lino, M.
2001 *Expenditures on Children by Families.* Available: http://www.usda.gov/cnpp/FENR/FENRv14n2/fenrv14n2p3.pdf [accessed August 23, 2004].

Love, J.M., L. Harrison, A. Sagi-Schwartz, M.H. van Ijsendoorn, C. Ross, J.A. Ungerer, H. Raikes, C. Brady-Smith, K. Boller, J. Brooks-Gunn, J. Constantine, E.E. Kisker, D. Paulsell, and R. Chazan-Cohen
2003 Child care quality matters: How conclusions may vary with context. *Child Development* 74(4):1021-1033.

Lundberg, S., and R. Pollak
1996 Bargaining and distribution in marriage. *Journal of Economic Perspectives* 10(Fall):139-158.

Lundberg S.J, and R. Starz
2000 Inequality and race: Models and policy. In *Meritocracy and Economic Inequality*, K. Arrow, S. Bowles, and S. Durlauf, eds. Princeton, NJ: Princeton University Press.

Mäler, K-G.
1996 *Resource Accounting, Sustainable Development and Well-Being.* Beijer Discussion Paper Series No. 75. Stockholm: The Royal Swedish Academy of Sciences.

Mankiw, N.G., D. Romer, and D.N. Weil
1992 A contribution to the empirics of economic growth. *Quarterly Journal of Economics* 107(2):407-437.

Mayerhauser, N., S. Smith, and D.F. Sullivan
2003 Preview of the 2003 comprehensive revision of the National Income and Product Accounts. *Survey of Current Business* (August):7-31.

McCabe, L.A., M. Cunnington, M., and J. Brooks-Gunn
2004 The development of self-regulation in young children: Individual characteristics and environmental contexts. In *Handbook of Self-Regulation: Research, Theory, and Application,* R.F. Baumeister and K.D. Vohs, eds. New York: Guilford Publications.

McKeown, T.
1976 *The Role of Medicine: Dream Mirage, or Nemesis?* London: Nuffield Provincial Hospitals Trust.

Mead, C., C. McCully, and M. Reinsdorf
2003 Income and outlays of households and of nonprofit institutions serving households. *Survey of Current Business* (April):13-17.

Michael, R.
1996 Money illusion: The importance of household time use in social policy making. *Journal of Family and Economic Issues* 17:245-260.

Moss, M.
1973 *The Measurement of Economic and Social Performance*. Studies in Income and Wealth, Volume 38. New York: National Bureau of Economic Research.

Moulton, B.
2002 Presenting Imputations in the National Income and Product Accounts. Paper presented at the OECD Meeting of National Accounts Experts, STD/NA(2002)14, October 8-11. Bureau of Economic Analysis, U.S. Department of Commerce.

Murnane, R., J. Willet, and F. Levy
1995 The growing importance of cognitive skills in wage determination. *Review of Economics and Statistics* 77:251-266.

Murphy, K., and R. Topel
2003 *Measuring the Gains from Medical Research*. Chicago: University of Chicago Press.

Murphy, M.
1978 The value of nonmarket housework production: Opportunity cost versus market cost estimates. *Review of Income and Wealth* 24(3):243-255.
1982 Conceptual estimates of the value of household work in the United States for 1976. *Review of Income and Wealth* 28(1):29-43.

Murray, C.J., and A.D. Lopez
1996 *The Global Burden of Disease*. Geneva, Switzerland: World Health Organization, Harvard School of Public Health, and The World Bank.

National Bureau of Economic Research
1978 *The Measurement of Economic and Social Performance*. Draft report to the National Science Foundation. Cambridge, MA: National Bureau of Economic Research.

National Institute of Child Health and Human Development Early Child Care Research Network
1997 The effects of infant child care on infant-mother attachment security: Results of the NICHD study of early child care. *Child Development* 68:860-879.

National Research Council
1998 *Measuring the Government Sector of the U.S. Economic Accounts*. Committee on National Statistics, C.M. Slater and M.H. David, eds. Commission on Behavioral and Social Science and Education. Washington, DC: National Academy Press.
1999 *Nature's Numbers: Expanding the National Economic Accounts to Include the Environment*. Panel on Integrated Environmental and Economic Accounting, W.D. Nordhaus and E.C. Kokkelenberg, eds. Committee on National Statistics, Commission on Behavioral and Social Science and Education. Washington, DC: National Academy Press.
2000 *From Neurons to Neighborhoods: The Science of Early Childhood Development*. Committee on Integrating the Science of Early Childhood Development, J.P. Shonkoff, and D.A. Phillips, eds. Board on Children, Youth, and Families, Division of Behavioral and Social Sciences and Education and Institute of Medicine. Washington, DC: National Academy Press.
2002a *At What Price? Conceptualizing and Measuring Cost-of-Living and Price Indexes*. Panel on Conceptual, Measurement, and Other Statistical Issues in Developing Cost-of-Living Indexes, C.L. Schultze and C. Mackie, eds. Committee on National Statistics, Division of Behavioral and Social Sciences and Education. Washington, DC: National Academy Press.
2002b *Estimating the Public Health Benefits of Proposed Air Pollution Regulations*. Committee on Estimating the Health-Risk-Reduction Benefits of Proposed Air Pollution Regulations. Board on Environmental Studies and Toxicology, Division of Earth and Life Sciences. Washington, DC: National Academy Press.

Neal, D., and W. Johnson
1996 The role of premarket factors in black-white wage differentials. *Journal of Political Economy* 104:869-895.

Nelson, R.R., and E. Phelps
1966 Investment in humans, technological diffusion, and economic growth. *American Economic Review* (May):56, 69-75.

Niemi, I., and E. Hamunen
1999 Developing a satellite account for household production. *Bulletin of the International Statistical Institute, Proceedings of the 52nd Session*. Available: http://www.stat.fi/isi99/proceedings.html [accessed January 28, 2004].

Nordhaus, W.D.
2003 The health of nations: The contribution of improved health to living standards. In *Measuring the Gains from Medical Research*, K. Murphy and R. Topel, eds. Chicago: University of Chicago Press.

2004 Principles of National Accounting for Non-Market Accounts. Paper presented at the CRIW Architecture for the National Accounts Conference, April 16-17, Washington, DC, Department of Economics, Yale University.

Nordhaus, W., and J. Tobin

1972 Is growth obsolete? In *Economic Growth*. National Bureau of Economic Research General Series #96E. New York: Columbia University Press.

Office of the Inspector General, U.S. Department of Health and Human Services

1998 *Hospital Stays for Medicare Beneficiaries Who Are Discharged to Home Health Agencies.* Available: http://oig.hhs.gov/oei/reports/oei-02-94-00321.pdf [accessed October 13, 2004].

Organisation for Economic Co-operation and Development

2000 Literacy in the Information Age : Final Report of the International Adult Literacy Survey. Paris, France: Organisation for Economic Co-operation and Development.

2002 *Measuring the Non-Observed Economy: A Handbook.* Paris, France: Organisation for Economic Co-operation and Development.

Oser, J., and S.L. Brue

1988 *The Evolution of Economic Thought*, 3rd edition. Orlando, FL: Harcourt Brace.

Palmquist, R.B., and A. Israngkura

1999 Valuing air quality with hedonic and discrete choice models. *American Journal of Agricultural Economics* 81:1128-1133.

Pamuk E., D. Makuc, K. Heck, C. Reuben, K. Lochner

1998 *Socioeconomic Status and Health Chartbook. Health, United States, 1998.* Hyattsville, MD: National Center for Health Statistics.

Peskin, H.M.

1999 Alternative resource and environmental accounting approaches and their contribution to policy. In *Resources Accounting in China*, A. Lanza, ed. New York: Kluwer Academic Publishers.

Peskin, H.M., and M.S. Delos Angeles

2001 Accounting for environmental services: Contrasting the SEEA and ENRAP approaches. *Review of Income and Wealth* 47(2):203-220.

Peskin, H.M., and E. Lutz.

1990 *A Survey of Resource and Environmental Accounting in Industrialized Countries.* Environment Department Working Paper No. 37. Washington, DC: The World Bank.

Pigou, A.C.

1920 *The Economics of Welfare.* London: Macmillan Press.

Plomin, R., J.C. Defries, I.W. Craig, P.R. McGuffin,

2002 *Behavioral Genetics in the Postgenomic Era.* Washington D.C.: American Psychological Association.

Policy Research Initiative

2003 *Social Capital: Building on a Network-Based Approach.* Available: http://policyresearch.gc.ca/page.asp?pagenm=Social_Capital_BNBA [accessed August 23, 2004].

Pollak, R.

1989 *The Theory of the Cost-of-Living Index.* New York: Oxford University Press.

1999 Notes on time use. *Monthly Labor Review.* 12(August):7-11.

Popkin, J.

2000 Data watch: The U.S. national income and product accounts. *Journal of Economic Perspectives* 14(Spring):205-213.

Portney, P.

1994 The contingent valuation debate: Why economists should care. *Journal of Economic Perspectives* 8(Fall):3-18.

Preston, S.
1993 Demographic Change in the United States, 1970-2000. In A.M. Rappaport and S. Scheiber, eds. *Demography and Retirement: The Twenty-First Century.* Westport, CT: Praeger Pubs.

Preston, S., and M. Haines
1991 *Fatal Years: Child Mortality in Late Nineteenth-Century America.* Princeton: Princeton University Press.

Putnam, R.D.
2001 *Bowling Alone: The Collapse and Revival of American Community.* New York: Simon and Schuster.

Reid, M.
1934 *The Economics of Household Production.* New York: Wiley.

Reilly, M.C., A.S. Zbrozek, and E.M. Dukes
1993 The validity and reproducibility of a work productivity and activity impairment instrument. *Pharmacoeconomics* 4(5):353-365.

Repetto, R., W. Magrath, M. Wells, C. Beer, and F. Rossini
1989 *Wasting Assets: Natural Resources in the National Income Accounts.* Washington, DC: World Resources Institute.

Rice, D., B. Cooper, and R. Gibson
1982 U.S. national health accounts: Historical perspectives, current issues, and future projections. In *La Sante Fait Ses Comptes (Accounting for Health)*, E. Levy, ed. Paris, France: Economica.

Riley, J.C.
2001 *Rising Life Expectancy: A Global History.* London: Cambridge University Press.

Robinson, J., and G. Godbey
1999 *Time for Life: The Surprising Ways Americans Use Their Time*, 2nd ed. University Park, PA: Pennsylvania State University Press.

Romer, P.M.
1990 Endogenous technical change. *Journal of Political Economy* 98:S71-S102.

Ruggles, R.
1983 The United States National Income Accounts, 1947-77: Their conceptual basis and evolution. In *The U.S. National Income and Product Accounts: Selected Topic,* M.F. Foss, ed. Chicago: University of Chicago Press.

Ruggles, R., and N.D. Ruggles
1982 Integrated economic accounts for the United States 1947-1980. *Survey of Current Business* (May):1-53.

Sandberg, J.F., and S. L. Hofferth
2001 Changes in children's time with parents, U.S. 1981-1997. *Demography* 38(3):423-436.

Sayer, L.C., S.M. Bianchi, and J.P. Robinson
2003 Are parents investing less in children? Trends in mothers' and fathers' time with children. Manuscript, revised version of a paper presented at the American Sociological Association Annual Meeting, August, 2000. Department of Sociology, Ohio State University.

Schwartz, L.K.
2002 The American Time Use Survey: Cognitive pretesting. *Monthly Labor Review* 125(February):34-44.

Shapiro, M.D., and D.W. Wilcox
1996 Mismeasurement in the consumer price index: An evaluation. In *NBER Macroeconomics Annual 1996*, B.S. Bernanke and J.J. Rotemberg, eds. Cambridge, MA: MIT Press.

Simon, A.
1990 A mechanism for social selection and successful altruism. *Science.* New Series 250(4988):1665-1668.

Smith, A.
1776 *The Wealth of Nations*. Amherst, NY: Prometheus Books.
Smith, J.
2004 *Consequences and Predictors of New Health Events*. NBER Working Paper No. 10063. Cambridge, MA: National Bureau of Economic Research.
Smith, V.K., and J.C. Huang
1995 Can markets value air quality? A meta-analysis of hedonic property value models. *Journal of Political Economy* 103:209-225.
Sorokin, P., and C. Berger
1939 *Time-Budgets of Human Behavior*. Cambridge, MA: Harvard University Press.
Spence, A.M.
1974 *Market Signaling: Informational Transfer in Hiring and Related Screening Processes*. Cambridge, MA: Harvard University Press.
Stafford, F.P.
1996 Early education of children in families and schools. In *Household and Family Economics*, P.L. Menchik, ed. Boston: Kluwer Academic Publishers.
Statistics Canada
1995 *Households' Unpaid Work: Measurement and Valuation*. Studies in National Accounting. Ottawa: Statistics Canada.
Strauss, J., and D. Thomas
1998 Health, nutrition, and economic development. *Journal of Economic Literature* 36(2):766-817.
Sturm, R., C.R. Gresenz, R.L. Pacula, and K.B. Wells
1999 Labor force participation by persons with mental illness. *Psychiatric Services* 50(11):1407.
Szalai, S.
1973 *The Use of Time*. The Hague, The Netherlands: Mouton.
Torrance, G.W.
1987 Utility approach to measuring health-related quality of life. *Journal of Chronic Diseases* 40:6:593-600.
Triplett, J.E.
1999 *Measuring the Prices of Medical Treatments*. Washington, DC: Brookings Institution Press.
2001 Measuring Health Output: The Draft Eurostat Handbook on Price and Volume Measures in National Accounts. Paper presented at the Eurostat-CBS Seminar, Voorburg, Netherlands, March 14-16. The Brookings Institution.
2002 Integrating Cost-of-Disease Studies into Purchasing Power Parities (PPP). Unpublished paper prepared for OECD Working Party on Social Policy Workshop What is Best and at What Cost? OECD Study on Cross-National Differences on Aging-Related Diseases, Paris, June 17.
United Nations
1993 *Handbook of National Accounting: Integrated Environmental and Economic Accounting*, Series F, No. 61. New York: U.N. Department for Economic and Social Information and Policy Analysis, Statistical Division.
2003a *Handbook on Nonprofit Institutions in the System of National Accounts*. New York: United Nations.
2003b *Handbook of National Accounting: Integrated Environmental and Economic Accounting 2003*. Available: http://unstats.un.org/unsd/environment/seea2003.pdf [accessed May 12, 2004].
Urban Institute and National Center for Charitable Statistics
2000 National Center for Charitable Statistics Data Files. Available: http://nccsdataweb.urban.org/FAQ/index.php?category=90 [accessed October 12, 2004].

U.S. Bureau of Labor Statistics

1997 *Measurement Issues in the Consumer Price Index.* Paper prepared at the request of Rep. Jim Saxton, Chairman, Joint Economic Committee, to respond to the Final Report of the Advisory Commission to Study the Consumer Price Index. Available: http://www.bls.gov/cpi/cpigm697.htm [accessed October 12, 2004].

2003 *Volunteering in the United States, 2003.* Available: http://www.bls.gov/news.release/volun.nr0.htm [accessed August 23, 2004].

U.S. Census Bureau

1999 *Statistical Abstract of the United States: 1999*, 119th edition. Washington, DC; U.S. Department of Commerce.

2002 *Statistical Abstract of the United States: 2002*, 122nd edition. Washington, DC: U.S. Department of Commerce.

2004 *Housing Vacancies and Homeownership, Annual Statistics: 2004.* Available: http://www.census.gov/hhes/www/housing/hvs/historic/histt14.html [accessed August 23, 2004].

U.S. Department of Health and Human Services

2000 *Healthy People 2010: Understanding and Improving Health, 2nd ed.* Washington, DC: U.S. Government Printing Office.

U.S. General Accounting Office

1995 *Long Term Care: Current Issues and Future Directions.* GAO/HEHS-95-109. Washington, DC: U.S. Government Printing Office.

2003 *Vehicle Donations: Benefits to Charities and Donors, But Limited Program Oversight.* GAO-04-73. Washington, DC: General Accounting Office U.S. Government Printing Office.

U.S. Office of Management and Budget

2003 *Budget of the United States, 2003.* Available: http://www.whitehouse.gov/omb/budget/fy2003 [accessed May 12, 2004].

van de Ven, P., B. Kazeimer, and S. Keuning

1999 Measuring Well-Being with an Integrated System of Economic and Social Accounts. Unpublished working paper, Statistics Netherlands.

Viscusi, W.K., and J.E. Aldy

2003 *The Value of a Statistical Life: A Critical Review of Market Estimates Throughout the World.* NBER Working Paper 9487. Cambridge, MA: National Bureau of Economic Research.

Walsh, J.R.

1935 Capital concept applied to man. *Quarterly Journal of Economics* 49(2):255-285.

Weimer, D., and M. Wolkoff

2001 School performance and housing values: Using non-contiguous district and incorporation boundaries to identify school effects. *National Tax Journal* 54(June):231-253.

Weinstein M.C., J.E. Siegel, M.R. Gold, M.S. Kamlet, and L.B.Russell

1996 Recommendations of the Panel on Cost-Effectiveness in Health and Medicine. *Journal of the American Medical Association* 276:1253-1258.

Weisbrod, B.A.

1975 Toward a theory of the nonprofit sector in a three-sector economy. In *Altruism, Morality, and Economic Theory*, E. Phelps, ed. New York: Russell Sage Foundation.

Weitzman, M.

1976 On the welfare significance of national product in a dynamic economy. *Quarterly Journal of Economics* 90:156-162.

Wilson, W.J.

1995 *The Truly Disadvantaged: The Inner City, the Underclass and Public Policy.* Chicago: University of Chicago Press.

Wise, P.
2004 Defining positive health. Presentation to the National Children's Study Advisory Committee, Washington, DC, March 4. Department of Pediatrics, Boston University School of Medicine.

Wolf, A.
2002 *Does Education Matter? Myths About Education and Economic Growth*. New York: Penguin Books.

Wolfe, B., and R. Haveman
2001 Accounting for the social and non-market benefits of education. In *The Contribution of Human and Social Capital to Sustained Economic Growth and Well-Being*, J. Helliwell, ed. OECD/Human Resources Development Canada. Vancouver: University of British Columbia Press.
2002 Social and nonmarket benefits from education in an advanced economy. In *Proceedings from the Federal Reserve Bank of Boston's 47th Economic Conference*. Boston: Federal Reserve Bank of Boston.

Young, P.C.
1993 Nonprofit Institutions in an Input-Output Framework. *Voluntas* 4:465–485.

Appendix

Biographical Sketches of Panel Members and Staff

Katharine G. Abraham (*Chair*) is professor of survey methodology and adjunct professor of economics at the University of Maryland and a research associate at the National Bureau of Economic Research. She was commissioner of the Bureau of Labor Statistics from 1993 to 2001. Prior to her government service, she taught at the University of Maryland and the Sloan School of Management at the Massachusetts Institute of Technology and was a research associate at the Brookings Institution. Her research includes work on economic measurement issues, together with studies of internal labor markets and comparitive labor market analyses. She has been an associate editor of the *Quarterly Journal of Economics* and an assistant editor of the *Brookings Papers on Economic Activity* and is an elected fellow of the American Statistical Association. She received a Ph.D. in economics from Harvard University and a B.S. in economics from Iowa State University, which in 2002 awarded her an honorary doctorate.

David Cutler is a professor in the Economics Department and in the Kennedy School of Government at Harvard University, and he is also a research associate at the National Bureau of Economic Research. During 1993 he was on leave to serve as senior staff economist at the Council of Economic Advisers and the National Economic Council. His research is concentrated in health economics, including measuring the health of the population and understanding how medical and nonmedical factors influence health. He is coeditor of the *Journal of Health Economics* and associate editor of the *Journal of Public Economics* and the *Journal of Economic Perspectives*. He is an elected member of the Institute of Medicine. He has been a member of numerous commissions and advisory groups, including the Social Security Advisory Council Technical Panel on Assumptions and Methods and the Medicare Technical Review Advisory Panel. He received a B.A., summa cum laude, from Harvard College and a Ph.D. in economics from the Massachusetts Institute of Technology.

Nancy Folbre is a professor in the Department of Economics at the University of Massachusetts and an associate editor of *Feminist Economics* at Amherst. Her work explores the interface between feminist theory and political economy, with a particular focus on caring labor and other forms of nonmarket work. Her research overlaps the fields of economic history, development, and social and family policy. She is the author of *Who Pays for the Kids? Gender and the Structures of Constraint* (Routledge, 1994) and *The Invisible Heart: Economics and Family Values* (New Press, 2001) and recently coedited a book with Michael Bittman, *Family Work: The Social Organization of Care* (Routledge, 1994). She received B.S. and M.A. degrees from the University of Texas and a Ph.D. in economics from the University of Massachusetts.

Barbara Fraumeni is chief economist at the Bureau of Economic Analysis. Previously, she was a professor of economics at Northeastern University, and she was also a research fellow of the Program on Technology and Economic Policy at the John F. Kennedy School of Government at Harvard University. Her areas of expertise and research interests include measurement issues and national income accounting; human and nonhuman capital, productivity, and economic growth; market and nonmarket accounts; investment in education, research, and development; and measurement of highway capital stock and the real output of government by function. She is a coauthor of *Productivity and U.S. Economic Growth* with Dale W. Jorgenson and Frank M. Gollop. She received a B.A. from Wellesley College and a Ph.D. from Boston College.

Robert E. Hall is the McNeil joint senior fellow at the Hoover Institution and a professor in the economics department, both at Stanford University. His research interests are in the behavior of the aggregate American economy, including the labor market, investment, and the stock market. He serves as director of the research program on economic fluctuations and growth of the National Bureau of Economic Research. He is a member of the National Academy of Sciences and a fellow of the American Academy of Arts and Sciences and of the Econometric Society. He received a B.A. from the University of California, Berkeley, and a Ph.D. from the Massachusetts Institute of Technology.

Daniel S. Hamermesh is Edward Everett Hale Centennial professor of economics at the University of Texas at Austin. He has taught at Princeton University and Michigan State University and has held visiting professorships at universities in the United States, Europe, Australia, and Asia. He is a fellow of the Econometric

Society, a research associate of the National Bureau of Economic Research and of the Institute for the Study of Labor, and past president of the Society of Labor Economists and of the Midwest Economics Association. He authored *Labor Demand, The Economics of Work and Pay* and a wide array of articles in labor economics. His research concentrates on labor demand, time use, and unusual applications of labor economics to suicide, sleep, and beauty. He received an A.B. from the University of Chicago and a Ph.D. from Yale.

Alan B. Krueger is the Bendheim Professor of Economics and Public Affairs at the Woodrow Wilson School and the Economics Department at Princeton University. His primary research and teaching interests are in the areas of labor economics, education, industrial relations, and social insurance. He is the author of *Education Matters: A Selection of Essays on Education* and the coeditor of the *Journal of Economic Perspectives.* His current research projects include a study of the effect of economic growth on employment and income of less skilled workers, an examination of the determinants of participation in terrorism, a study of the relationship between school quality and student outcomes, and the development of a new approach to measuring well-being based on time allocation. He writes a monthly column on economics for *The New York Times.* He is a fellow of the Econometric Society and the American Academy of Arts and Sciences. He was awarded the Kershaw Prize by the Association for Public Policy and Management in 1997 and the Mahalanobis Prize by the Indian Econometric Society in 2000. He received a B.S. from Cornell University and a Ph.D. from Harvard University.

Christopher Mackie (*Study Director*) is with the Committee on National Statistics (CNSTAT). In addition to working with this panel, he is working with the Panel on Accessing Data Access and Confidentiality, and he served as study director of the Panel on Conceptual, Measurement, and Other Statistical Issues in Developing Cost-of-Living Indexes. Prior to joining CNSTAT, he was a senior economist with SAG Corporation, where he conducted a variety of econometric studies in the areas of labor and personnel economics, primarily for federal agencies. He is the author of *Canonizing Economic Theory.* He has a Ph.D. in economics from the University of North Carolina; while a graduate student, he held teaching positions at the University of North Carolina, North Carolina State University, and Tulane University.

Robert Michael is the Eliakim Hastings Moore distinguished service professor in the Irving B. Harris Graduate School of Public Policy Studies at the University of Chicago. Previously, he was director of the National Opinion Research Center (NORC) and director of the West Coast office of the National Bureau of Economic Research. He also previously taught at Stanford University and the University of California at Los Angeles. His research interest and publications cover family economics, including the causes of divorce; the reasons for the growth of one-person households; the effects of inflation on families; and the consequences of the rise in women's employment for the family, especially children; and expenditure patterns in the household, including the factors that determine parental spending on children in various types of households. He serves on the boards of the Chapin Hall Center for Children and NORC and cochairs the Board of Visitors of Western Reserve Academy in Hudson, Ohio. He is an elected fellow of the American Association for the Advancement of Science.

Henry M. Peskin is president of Edgevale Associates, a consulting company. Formerly, he served on the staffs of the National Bureau of Economic Research, the Urban Institute, the Institute for Defense Analysis, and, most recently, Resources for the Future. With training in chemical engineering at the Massachusetts Institute of Technology, an undergraduate degree in political science from Wesleyan University, and a Ph.D. in economics from Princeton University, he has written extensively on methods to expand the national economic accounts in order to better measure resource and environmental degradation. As a consultant to the World Bank and the U.S. Agency for International Development, he has surveyed environmental accounting practices in industrialized countries and has advised developing countries on the design and implementation of environmental accounting systems. He was a member of the National Research Council's Panel on Integrated Environmental and Economic Accounting.

Matthew D. Shapiro is the Lawrence R. Klein Collegiate Professor of Economics and research professor at the Survey Research Center of the University of Michigan, and he is also a research associate of the National Bureau of Economic Research. His general area of research is macroeconomics, and he has carried out projects on investment and capital utilization, business-cycle fluctuations, consumption and saving, financial markets, monetary and fiscal policy, time-series econometrics, and improving national economic statistics. During 1993-1994 he served as senior economist at the Council of Economic Advisers. He is now a member of the Academic Advisory Panel of the Federal Reserve Bank of Chicago, the Federal Economic Statistics Advisory Committee, and the executive committee of the Conference on Research in Income and Wealth. He has been a member of the Committee on National Statistics. He received B.A. and M.A. degrees from Yale and a Ph.D. from the Massachusetts Institute of Technology.

Burton A. Weisbrod is the John Evans professor of economics and a faculty fellow in the Institute for Policy Research, at Northwestern University. He has served on the President's Council of Economic Advisers as a senior staff economist, on the National Institutes of Health National Advisory Research Resources Council, and on numerous advisory committees. He currently serves on the Users Advisory Committee of the Statistics of Income Division of the Internal Revenue

Service and as chair of the Social Science Research Council's Committee on Philanthropy and the Nonprofit Sector. His elected positions include president of the Midwest Economics Association, member of the executive committee of the American Economics Association, fellow and council member of the American Association for the Advancement of Science, and the Institute of Medicine. He has lectured widely in Asia, Australia, and Europe. His research focuses on technological change in medical care and on organization behavior in such industries as medical care, higher education, and the arts, where nonprofit and for-profit organizations coexist. His books include *To Profit or Not to Profit: the Commercial Transformation of the Nonprofit Sector* and *The Nonprofit Economy.* He holds a B.S. from the University of Illinois and a Ph.D. from Northwestern University.

Index

A

ABS. *See* Australian Bureau of Statistics
Accounting and data foundations, 39–54
  demographic data, 52–54
  measuring time use, 43–52
  overview of the national income and product accounts, 40–43
Accounting approaches, in the current environment, 169–171
Accounts
  balance-sheet, 40
  expanded set of, 2
  experimental, 2–3, 20
  income and product, 40–43
  input-output, 40
  national, 56
  satellite, 2–5, 11, 16–18
AFQT. *See* Armed Forces Qualification Test
Aggregate output, 26
Aggregate production, 69
Aggregate welfare measurement, 156
Amenity value, 31
American Association of Fundraising Counsel, Trust for Philanthropy, 159
American Community Survey, 7, 53
American Time Use Survey (ATUS), 7, 10, 19, 27, 39, 45–52, 71, 76, 88, 100, 126, 140, 147–148, 160
  recommendations for, 7, 46–48, 77
Armed Forces Qualification Test (AFQT), 113
Assignment of prices, 28–34
ATUS. *See* American Time Use Survey
Australia, 45, 76, 122, 154
Australian Bureau of Statistics (ABS), 45

B

Balance-sheet accounts, 40
Bargaining power, 60
BEA. *See* Bureau of Economic Analysis
Belgium, 154
Benefit-cost analysis, 157
BLS. *See* Bureau of Labor Statistics
Bookkeeping, implications of double-entry, 24–25
British National Child Development Study, 112
Bureau of Economic Analysis (BEA), 5, 15–16, 19, 21, 26, 43, 77, 94–95, 97, 99, 142–143, 154, 159, 161, 166–167
  recommendations for, 59
Bureau of Labor Statistics (BLS), 11, 19, 28, 45–52, 77, 88, 94, 100, 118, 119n, 146, 151, 154
  recommendations for, 7, 48–52
Bypass surgery, 131

C

Cable television, 90
Canada, 45, 76, 80, 154
Capital-market constraints, 24
Capital service flows, 99
Capital stock, 94, 171
Care services, 3
  noncompensated, 126
Census Bureau, 7, 28, 53, 94, 102
Centers for Medicare and Medicaid Services, 119
Chain-weighted quantity indexes, 34
Child care, 58, 64, 66, 81, 90, 105
Child Development Supplement of the Panel Study of Income Dynamics, 89
Children
  commitment to bear and raise, 92
  as human capital, family inputs to the development of, 88–90
  time use by, 90
Cholesterol intake, 129
CIPSEA. *See* Confidential Information Protection and Statistical Efficiency Act of 2002
Citizenship, personal habits of, 87
Civility, 87
Classification of goods and services, 25–26
Cleaning services, 19
Cleanliness, 61–62. *See also* Home cleanliness; Outdoor cleaning
Cognitive skills, 115
  current, 109
Collective goods, 141
Comparability, across countries, 16
Competitive pressures, 24
Compulsory schooling requirements, 106
Computers, 57, 89
Conceptual framework
  for the family's role in the production of human capital, 79–81
  for the government and private nonprofit sectors, 143–146
Conceptual framework for education, 94–97
  education inputs and outputs, market and nonmarket, 96
Conceptual framework for health, 119–125
  broader approach, 122–125
  health inputs and outputs of market and nonmarket, 124
  market-oriented approaches, 120–122
  national health expenditures, 121
Conceptual issues, 5–6, 23–37
  assigning prices, 28–34
  classifying goods and services, 25–26
  counting and valuation issues, 34–36
  externalities, 26–27
  implications of double-entry bookkeeping, 24–25
  marginal and total valuation, 36–37
  measuring quantities, 27–28
Confidential Information Protection and Statistical Efficiency Act of 2002 (CIPSEA), 159
Constructed health expenditure accounts, 13
Consumer durables, 72
Consumer Expenditure Survey, 52, 91
Consumer Price Index (CPI), 77, 119n
Consumption value, 71
Cost-of-disease approach, treatment-based, 122
Counting issues, 34–36
CPI. *See* Consumer Price Index
CPS. *See* Current Population Survey
Cross-country comparisons, 110
Cross-household variation, 60
Current Population Survey (CPS), 46–48, 51, 53, 126, 146–147, 160

D

Data needs, 6–8
  for future environmental directions, 174–175
  in the government and private nonprofit sectors, 159–160
  in health, 140
Data needs in home production, 76–78
  input quantities and prices (time use), 76–77
  output quantities and prices, 77–78
Day care, 83
Death, causes of, 124
Degradation, valuing, 169
Demographic data, 7, 27–28, 52–54, 130–131
  recommendations for, 7–8, 53–54
Department of Agriculture, 91
Department of Commerce, 19
Department of Defense, 145
Department of Health and Human Services, 136
  Centers for Medicare and Medicaid Services, 119, 136n
  National Health Accounts, 120
Department of Labor, 19

Depreciation, of natural resources, 165, 167
Development of children's human capital, family inputs to, 88–90
Developmental psychology, 82
Direct quantity-based index, 118
Disamenity costs, 116n
Discrete choice model, 177
Disease-adjusted life years, 136
"Disease state" approach, 3, 134
Divorce rates, 79
Donated goods
  in the government and private nonprofit sectors, 153
  recommendations for, 146
Double-entry bookkeeping, 6, 26, 76, 161
  implications of, 24–25
Down's syndrome, 90
Drinking, 129
Drugs, new, 118, 120
DVD players, 90

E

Early Child Care Research Network, 83
Earnings, education's link to higher, 106
Econometrics, 132
Economic inequality, 13
Economic production, 1
Economic theory, 29
Economically valued nonmarket factors, 10
Education, 3, 80, 93–116
  conceptual framework for, 94–97
  inputs and outputs, market and nonmarket, 96
  link to higher earnings, 106
  measuring and valuing inputs in, 97–105
  measuring and valuing output in, 105–116
Education satellite account, recommendations for, 97
Educational psychology, 84
Educational satellite accounts, 93
Empathy, 87
ENRAP. *See* Environmental and Natural Resources Accounting Project
Environment, 4, 130, 163–177
  current accounting approaches, 164, 169–171
  definition and scope of coverage, 164–169
  future directions, 171–175
  the social environment, 175–177
Environmental and Natural Resources Accounting Project (ENRAP), 170, 172–173
Environmental Protection Agency (EPA), 164, 172–173
Environmental resources, renewable, in defining environment, 167–169
Estimation approaches and practice, in future environmental directions, 171–173
Ethical conduct, value of, 87
Euroqual-5D (five domains), 136
Eurostat, 56
  Handbook, 121
Experimental methods, 2–3, 20
Externalities, 26–27

F

"Family disintegration," 18
Family role in the production of human capital, 4, 79–92
  conceptual framework, 79–81
  defining human capital, 81–83
  family inputs to the development of children's human capital, 88–90
  the human capital production function, 83–88
  valuing the time parents devote to children, 91–92
Federal efforts, 97
  recommendations for, 164
Federal Trade Commission, 55
501(c)(3) organizations, 142, 145
501(c)(4) organizations, 142–143, 145
Food preparation, 89
Form 990, 159
Form 990T, 159
Fourth World Conference on Women, 44
"Free-rider" behavior, 157
Full capital service flow, 99
Future directions in environmental accounting, 171–175
  data needs, 174–175
  estimation approaches and practice, 171–173
  linkage with other nonmarket accounting efforts, 173–174

G

GAAP. *See* Generally accepted accounting principles
Gardeners, 72
GDI. *See* Gross domestic income
GDP. *See* Gross domestic product
General Accounting Office, 153
General Social Survey (Canada), 45
Generalist approach, 102
Generally accepted accounting principles (GAAP), 158, 161
Genetic differences, 84
Giving and Volunteering Survey, 146
Goods and services, classification of, 25–26
Government inputs
  and market inputs, 97–99
  recommendations for, 145
Government satellite accounts, 161
Government sector, 4, 141–162
  conceptual framework, 143–146
  data requirements, 159–160
  donated goods, 153
  measuring and valuing output in, 153–159
  volunteer labor, 146–152
Grand Canyon, 34, 130
Grants-in-aid, 97
Great Depression, 14
Griliches, Zvi, 54, 153
Gross domestic income (GDI), 40–43
Gross domestic product (GDP), 1–2, 12, 16, 19, 23, 34, 40–44, 56–58, 62, 73–74, 93, 97–99, 120–121, 143, 154, 166
  finding a replacement for, 16
Gross investment, 41
Grossman, Michael, 124

H

Handbook of National Accounting: Integrated Environmental and Economic Accounting, 163
Handbook on Nonprofit Institutions in the System of National Accounts, 121, 144, 150, 155
Head Start, 84
Health, 3–4, 117–140
  adverse shocks to, 132
  changes in, 125
  conceptual framework for, 119–125
  data requirements, 140
  improvements in, 123
  inputs and outputs of market and nonmarket, 124
  measuring and valuing, 131–140
  measuring and valuing inputs in, 125–131
  valuing increments of, 137–140
Health and Labor Questionnaire, 136n
Health capital, producing, 86
Health-impairment approach, 134
Health research expenditures, as a percentage of GDP, 121
Health satellite accounts, 125
  recommendations for, 118, 131
Health status, measuring, 133–137
Healthy activities, recommendations for, 128
Hedonic models, 114, 118, 174, 176–177
High School and Beyond Survey, 112
Home cleanliness, production function for, 61
Home-produced meals, in a household production account, stylized account for, 60
Home production, 3, 55–78
  data requirements, 76–78
  factory analogy, 59–62
  measuring and valuing output of, 74–76
  measuring inputs in, 63–68
  valuing inputs of, 68–73
Home production data needs, 76–78
  input quantities and prices (time use), 76–77
  output quantities and prices, 77–78
Home schooling, 64
Household as a factory, 59–62
  production function for home cleanliness, 61
  stylized account for home-produced meals in a household production account, 60
Household members' wage rates, in valuing inputs of home production, 69–70
Household production account
  home-produced meals in, 60
  recommendations for, 64, 74–76
Household technology, 60–61
Housing prices, 114
Housing value approach, to measuring and valuing output in education, 113–116
Human capital
  acquiring, 114
  defining, 81–83
  family inputs to the development of children's, 4, 88–90
  investment in, 10, 21, 81
  production function of, 83–88

I

IALS. *See* International Adult Literacy Survey
IALS prose scores, 110
IEESA, 166–167, 171
Immigration policies, 80
"Incentive-enhancing" preferences, 82
Income accounts
  national, 40–43
  overview of national, 40–43
Incremental earnings approach to measuring and valuing output in education, 111–113
Increments of health, valuing, 137–140
Independent Sector, 146
Index-number theorists, 34
Indexes
  chain-weighted quantity, 34
  direct quantity-based, 118
  price-deflated quantity, 118
  quality-of-life, 132, 176
Indicator approach to measuring and valuing output, 107–111
  IALS prose scores, 110
  NAEP mathematics scores, 109
  NAEP reading scores, 108
Infants, viewed as outputs, 80
Infectious diseases, 123
Input-output accounts, 40
Inputs
  in education, measuring and valuing, 97–105
  in health, measuring and valuing, 125–131
  in home production, measuring, 63–68
  of home production, valuing, 68–73
  quantities and prices (time use) for home production data requirements, 76–77
Internal Revenue Code, 143, 145
International Adult Literacy Survey (IALS), 110, 114–115
Internet, 57
Investment, 94, 101, 111
Investments families make, in preparing children for the future, 4–5
IRS Form 990, 158–159
Italy, 154

J

Journal of Economic Perspectives, 157n

K

"Knowledge sector," 120
Kuznets, Simon, 1, 9, 55

L

Labor supply, 139
Laissez faire societies, 168
Laundry example, 75, 77
Leisure, 19, 61
Leontief system, 170

M

Marginal valuation, and total valuation, 36–37
Marital matching, idiosyncrasies inherent in, 51
Market and nonmarket factors, in defining environment, 165–166
Market-based production technology, 59
Market inputs, and government inputs, 97–99
Market-oriented approaches, to health, 120–122
Market prices, 155–156
Market service providers, 31
Market substitutes, price of, in valuing inputs of home production, 68–69
Maternal employment, 89
Meal preparation, 29
Measure of Economic and Social Performance project (MESP), 173
Measurement objectives, 17–19
Measuring and valuing health, 131–140
  defining the output, 131–133
  measuring health status, 133–137
  valuing increments of health, 137–140
Measuring and valuing inputs in education, 97–105
  market and government inputs, 97–99
  national education expenses, 97–98
  nonmarket time inputs, 100–105
Measuring and valuing inputs in health, 125–131
Measuring and valuing output in education, 105–116
  housing value approach, 113–116
  incremental earnings approach, 111–113
  indicator approach, 107–111
Measuring and valuing output in government and private nonprofit sectors, 153–159
  value imputations, 158–159
  zero price problem, 155–158

Measuring and valuing output of home production, 74–76
Measuring health status, 133–137
Measuring inputs in home production, 63–68
  time spent in home production, 65, 67
Measuring quantities, 27–28
Measuring time use, 43–52
  American Time Use Survey, 45–52
  previous collections of time-use data, 44–45
  problems and nonproblems with ATUS, 48–52
Medical care, 3, 22, 80, 118, 123
  as a percentage of GDP, 121
Medicare program, 43
Medicine, discoveries in, 13
MESP. *See* Measure of Economic and Social Performance project
Microwave ovens, 57
Mill, John Stuart, 93
Monetary values, 35
Motivation, 12–14
Mozambique, 155
Multinational Time Use Survey, 89

N

NAMEA system, 170
National accounts, 56
National Assessment of Educational Progress (NAEP), 108–110, 114–115
  mathematics scores, 109
  reading scores, 108
National Bureau of Economic Research (NBER), 55, 173
National Center for Charitable Statistics, 159
National Center for Education Statistics (NCES), 52–53, 94, 97
National Center for Health Statistics (NCHS), 28, 122
National Education Association, 94
National education expenditures, 97–98
National health accounts, 120, 136n, 140
National Health and Nutrition Examination Survey (NHANES), 140
National health expenditures, 121
National Health Interview Survey, 140
National Health Interview Survey on Disability, 126
National income accountants, 75
National income and product accounts (NIPAs), 1–2, 5, 9–12, 16–17, 20–27, 39–43, 54–55, 63, 73, 78, 94, 117–122, 133, 142–148, 153, 156, 162, 166, 168–169, 172, 176
  gross domestic product and gross domestic income, 41
  imputations in, 42
  scope of coverage in, 14–16
National income trends, 149
National Institute of Child Health and Human Development, Early Child Care Research Network, 83
National Institutes of Health, 157
National Longitudinal Survey of Youth, 113
National monuments, 34
National Research Council, 35, 119n
National Science Foundation, 66, 157, 173
National well being, 16
Natural resource depreciation, 165, 167
Nature's Numbers, 4, 130, 163–164, 167, 171–175
NBER. *See* National Bureau of Economic Research
NCES. *See* National Center for Education Statistics
NCHS. *See* National Center for Health Statistics
"Near-market" activities, 18
Netherlands, 122, 155, 170
Netting-out rule, 129
New drugs, 118, 120
New Nonprofit Almanac, 142
NHANES. *See* National Health and Nutrition Examination Survey
NIPAs. *See* National income and product accounts
Noncompensated care services, 126
Nonlabor inputs, inputs of home production, valuing, 73
Nonmarket accounting priorities, 19–23
  linkage in future environmental directions, 173–174
Nonmarket factors, 2, 5–6, 10, 13, 35. *See also* Market and nonmarket factors
  development of, 12
  production, 21, 32
  recommendations for, 35, 37
  in service-oriented areas, 8
Nonmarket satellite accounts, recommendations for, 23

Nonmarket time inputs, 100–105
time estimates, 100
time valuations, 101–105
Nonmedical technology, 3
Nonprofit economic activity, 11
recommendations for, 145
Nonprofit organizations, 141–144
Nordhaus, William, 18
Nutrition, 123

O

OECD. *See* Organisation for Economic Co-operation and Development
Office for National Statistics (United Kingdom), 45, 56, 81
"On call" schedules, 88
Opportunity-cost-based approach, 6, 30, 102
Opportunity costs
calculating, 32
non-negative, 103
Organisation for Economic Co-operation and Development (OECD), 63n, 154
Aging-Related Diseases Study, 122
Out-of-pocket expenditures, 88
Outdoor cleaning, example of valuing inputs of home production, 72–73
Output in education
measuring and valuing, 105–116
recommendations for, 116
Output in government and private nonprofit sectors, measuring and valuing, 153–159
Output in measuring and valuing health, defining, 131–133
Output of home production, measuring and valuing, 74–76
Output quantities and prices, for home production data requirements, 77–78
Own-time inputs, 128
Owner-occupied housing, rental value of, 15
Ownership, 165

P

Paid caregivers, 91
Panel Study of Income Dynamics (PSID), 91
Child Development Supplement, 89
Panel to Study the Design of Nonmarket Accounts, 2, 10
Parenting, 85, 102
Personal Consumption Expenditure report, 77
Personal responsibility, 82, 87
Philippines, 155, 170, 172–173
Physical capital stock, 10
Pigou, A.C., 14
Policy, 124, 168
Policy makers, 1
Pollution, 28, 130, 164–167, 172, 175
Poverty lines, 13
Price-deflated quantity index, 118
Price indexes, 27
Prices
assignment of, 28–34
of market substitutes, in valuing inputs of home production, 68–69
Private nonprofit sector, 4, 141–162
conceptual framework, 143–146
data requirements, 159–160
donated goods, 153
measuring and valuing output in, 153–159
volunteer labor, 146–152
Pro bono legal services, 147, 149–150
"Pro-social" preferences, 82
Product accounts
national, 40–43
overview of national, 40–43
Productive capacity, 82
Productivity
economic, 1
estimating, 94
for home cleanliness, 61
in medical care, 117, 119
trends in, 115
Productivity-equivalent replacement wage, 102–103
PSID. *See* Panel Study of Income Dynamics
Public Broadcasting System, 155
Public goods, 141
Public policy, 79
Publication schedule, appropriate, 22

Q

QALYs. *See* Quality-adjusted life years
Quality-adjusted life-expectancy, 132
Quality-adjusted life years (QALYs), 124, 136
Quality-adjusted replacement cost, an alternative approach to valuing inputs of home production, 70–71
Quality-of-life indexes, 132, 176
Quantities, measuring, 27–28

Quantity-based index, direct, 118
QWB (quality of well-being) survey, 136

R

Rawl's veil of ignorance, 139
Recommendations
  for the American Time Use Survey, 7, 46–48, 77
  for the Bureau of Economic Analysis (BEA), 59
  for the Bureau of Labor Statistics, 7, 48–52
  for demographic data, 7–8, 53–54
  for donated goods, 146
  for an education satellite account, 97
  for federal efforts, 164
  for government inputs, 145
  for health satellite accounts, 118, 131
  for healthy activities, 128
  for a household production account, 64, 74–76
  for nonmarket satellite accounts, 23
  for nonprofit economic activity, 145
  for output in education, 116
  for a replacement cost measure, 6, 32
  for satellite accounts for household production, 73
  for statistical agencies, 4
  for time inputs, 68, 71
  for time inputs to education, 103–105
  for time use patterns, 100
  for unpaid time, 127
  for volunteer labor, 152
Regression discontinuity approach, 115
Reid, Margaret, 63
Religious organizations, 87, 143
Renewable environmental resources, in defining environment, 167–169
Replacement cost approach, recommendations for, 6, 32
Replacement wages, productivity-equivalent, 102–103
Research and development (R&D), 3, 130
Roofing example, 127

S

Safety devices, 3
Satellite accounts, 2–5, 11, 16–18
  for education, 93
  for government, 161
  for health, 125
  for household production, recommendations for, 73
  nonmarket, 23
School expenditures, 13
Scientific inventions, 13
Scope of coverage in the NIPAs, 14–16
SEEA. *See* System of integrated environmental and economic accounting
Self-maintenance, 112
Self-selection, 138
"Sentinel capabilities," 84
Services. *See* Goods and services
Sex act, 63–64
SF-36 questionnaire, 136
Shadow wages, 160
Sleep, 63–64, 112, 128–129
Smith, Adam, 93
Smoking, 129
SNA. *See* System of National Accounts
Social capital, 21, 34
Social scientists, encouraging in study of nonmarket activities, 10
Social welfare, contributions to, 1
"Soft skills," 82
South Africa, 155
Specialist approach, 102
Staged approach, 2
Standardized tests, 114
2003 Statistical Abstract, 13
Statistical agencies, recommendations for, 4
Statistics Canada, 88
Subsoil resources, in defining environment, 166–167
Sweden, 110, 155
Symptoms, grading, 136
System of integrated environmental and economic accounting (SEEA), 163, 170–171
System of National Accounts (SNA), 14, 17, 63n, 144, 151, 154, 169–171

T

Technical changes, 13, 57
Technology, 25, 29
  household, 60–61
  nonmedical, 130
Test scores, 107, 109, 113–115
Time diary, 46
Time estimates, 100

Time inputs, 4, 6–7, 14, 30
  market value of, 31
  recommendations for, 68, 71
Time inputs to education
  recommendations for, 103–105
  value of, 104
Time parents devote to children, valuing, 91–92
Time spent in home production, 65, 67
Time use
  by children, 90
  measuring, 43–52
  patterns of, recommendations for, 100
Time-use data, 39, 88
  previous collections of, 44–45
Time valuations, 101–105
Tobin, James, 18
Total valuation, and marginal valuation, 36–37
Transaction costs, 31, 60
Travel-cost method, 172, 174
Treatment-based cost-of-disease approach, 122
Trends
  historical, 13
  in national income, 149
  in productivity, 115
Trust for Philanthropy, 159

U

United Kingdom, 45, 56, 76, 81, 110, 122
United Nations, 14, 163
University of Essex, Multinational Time Use Survey, 89
University of Maryland, 44, 66
University of Michigan, 44, 50
Unpaid time, 11
  recommendations for, 127
Unpriced labor time, 148
Urban Institute, 146
  National Center for Charitable Statistics, 159
Utility-based measures, 135

V

Valuation issues, 34–36
Value
  imputations of, 158–159
  of life, 177
  of time inputs to education, 104
Valuing increments of health, 137–140
Valuing inputs of home production, 68–73
  example of outdoor cleaning, 72–73
  household members' wage rates, 69–70
  price of market substitutes, 68–69
  quality-adjusted replacement cost (alternative approach), 70–71
  valuing nonlabor inputs, 73
Valuing the time parents devote to children, 91–92
Vocational students, 114
Volunteer labor, 4, 96, 126, 144
  in the government and private nonprofit sectors, 146–152
  recommendations for, 152

W

Wage employment, women's participation in, 58
Wage rates, 82
  for household members, in valuing inputs of home production, 69–70
  marginal, 30
  shadow, 160
Walker, Francis, 55
Welfare lines, 13
Willingness-to-pay estimates, 174
Work Limitation Questionnaire, 136n
Work Productivity and Activity Impairment Questionnaire, 136n
World Wars, disruptions attributable to, 14

Y

YHL (years of healthy life expectancy), 136
Youth sports organization, 33

Z

Zero price problem, 155–158